Letort Paper

SHADES OF CORDS IN THE KUSH: THE FALSE HOPE OF "UNITY OF EFFORT" IN AMERICAN COUNTERINSURGENCY

Henry Nuzum

April 2010

*****

Comments pertaining to this report are invited and should be forwarded to: Director, Strategic Studies Institute, U.S. Army War College, 122 Forbes Ave, Carlisle, PA 17013-5244.

*****

This document was written in 2007 as an M.A. thesis for Johns Hopkins School of Advanced International Studies. Mr. Nuzum's project advisor for this thesis was Dr. Tom Keaney.

*****

All Strategic Studies Institute (SSI) publications are available on the SSI homepage for electronic dissemination free of charge. Hard copies of this report may also be ordered from our homepage free of charge. SSI's homepage address is: *www.StrategicStudiesInstitute.army.mil*.

*****

The Strategic Studies Institute publishes a monthly e-mail newsletter to update the national security community on the research of our analysts, recent and forthcoming publications, and upcoming conferences sponsored by the Institute. Each newsletter also provides a strategic commentary by one of our research analysts. If you are interested in receiving this newsletter, please subscribe on our homepage at *www.StrategicStudiesInstitute.army.mil/newsletter/*.

ISBN 1-58487-435-X

FOREWORD

Counterinsurgency (COIN) requires an integrated military, political, and economic program best developed by teams that field both civilians and soldiers. These units should operate with some independence but under a coherent command. In Vietnam, after several false starts, the United States developed an effective unified organization, Civil Operations and Revolutionary Development Support (CORDS), to guide the counterinsurgency. CORDS had three components absent from our efforts in Afghanistan today: (1) sufficient personnel (particularly civilian), (2) numerous teams, and (3) a single chain of command that united the separate COIN programs of the disparate American departments at the district, provincial, regional, and national levels. This monograph focuses on the third component, describing the benefits that unity of command at every level would bring to the American war in Afghanistan.

Section 1 sets forth a brief introduction to counterinsurgency theory, using a population-centric model, and examines how this warfare challenges the United States. Section 2 traces the evolution of the Provincial Reconstruction Teams (PRTs) and the country team; Section 3 describes problems at both levels. For comparison, in Section 4, the author examines similar efforts in Vietnam, where persistent executive attention finally integrated the government's counterinsurgency campaign under the unified command of the CORDS program. Section 5 discusses the American tendency towards a segregated response to cultural differences between the primary departments, executive neglect, and societal concepts of war. Section 6 argues that in its approach to COIN, the United

States has forsaken the military concept of unity of command in favor of "unity of effort" expressed in multiagency literature. Sections 7 and 8 describe how unified authority would improve our efforts in Afghanistan and propose a model for the future.

DOUGLAS C. LOVELACE, JR.
Director
Strategic Studies Institute

ABOUT THE AUTHOR

HENRY NUZUM works for Seacor Holdings, an American energy services company, in Dubai, United Arab Emirates. His previous assignments included the Office of Secretary of Defense (Policy), most recently as the Chief of Staff for the Office of Strategic Capabilities (2007-09). He worked briefly on the House Armed Services Committee. In 2005, he worked in Iraq for the International Republican Institute. Following college, Mr. Nuzum served as a boarding and strike officer aboard USS *John S. McCain* (DDG 56), completing two Arabian Gulf deployments. In addition to his sea service, he rowed in two Olympic Games and two World Championships through the Navy Sports Program. Mr. Nuzum earned an M.A. from Johns Hopkins School of Advanced International Studies, and an A.B. from Harvard University.

SUMMARY

The past 2 years have been the most violent of the Afghan insurgency thus far. Taliban and affiliates seek to undermine the state and sap the will of the occupying force. In response, the United States and the coalition pursue a counterinsurgency (COIN) campaign that coordinates military, political, and economic assistance to the Afghan government so that it may provide security and services to its people. If the effort succeeds, the government will win the confidence of the citizens, who will increasingly reject the insurgents.

To achieve this unified program at a subregional level, the United States has deployed civil-military Provincial Reconstruction Teams (PRTs) across the country. The collocation of different departmental representatives has improved the American response to insurgency. However, the program faces obstacles—too few civilians, too few teams, and multiple chains of command. This monograph examines the last aspect, the absence of a unified authority to guide American PRTs, and more briefly considers the management of our nationwide efforts.

Each PRT has nearly 100 uniformed members and two or three representatives of civilian agencies.[1] Guidance from Washington has divided the team's mission into three spheres: improving security, which falls to the military team leader; enhancing the capacity of the government in the provinces, the purview of the State Department officer; and facilitating reconstruction, the responsibility of the U.S. Agency for International Development (USAID) representative. The first team deployed in the fall of 2002; today, there are 25 PRTs in Afghanistan, a dozen of which are

American. While the teams have increased the standing of the government in the hinterlands, the absence of unified authority diminishes their impact. Because the military, State, and USAID personnel report through separate chains of command, performance depends on the relationships between departmental representatives. When personalities mesh, teams function well. However, because there is no on-site arbiter, and recourse to Kabul is convoluted, conflicts may fester without resolution.

Similar problems plague regional and national efforts. PRTs, led originally by Army Civil Affairs officers and now by Air Force and Navy officers, have uncertain influence over battalions led by combat arms officers. When the demands of reconstruction and traditional use of force compete, the Regional Commander (who directs maneuver battalions and the military elements of PRTs) arbitrates, often in favor of combat arms priorities. At the national level, General David Barno and Ambassador Zalmay Khalilzad established a tight relationship which improved civil-military coordination. However, the arrangement depended on those individuals. It was not an enduring construct and lapsed under their successors.

The United States is not new to this type of war. As Washington increased its commitment to South Vietnam through the mid-1960s, several departments directed segregated counterinsurgency programs. After numerous attempts failed to unify American COIN efforts, President Lyndon Johnson initiated the Civil Operations and Revolutionary Development Support program in 1967. CORDS assigned responsibility for counterinsurgency to the military and integrated all programs, including civilian, under its command. A 3-star equivalent civilian director, serving as a component commander of U.S. fighting forces in Vietnam, led the new organization. Civilians and uniformed members were interspersed throughout the organization and were vested with full authority over subordinates, no matter their parent department, agency, or service. Along with these organizational changes, the program dramatically increased the money and manpower devoted to counterinsurgency. It is true that the vast majority of the American military, which focused on conventional campaigns and assisting the South Vietnamese army, was excluded from CORDS's purview, as were national level civilian programs.[2] In spite of these shortcomings, the organization effectively integrated, within its parameters, the security, political, and economic portions of the COIN campaign from the district to national levels and contributed to the defeat of the Viet Cong insurgency.

Despite this success, the United States has neglected the lessons of Vietnam for at least three reasons. First, due to cultural differences, agencies resist integration. Second, the executive branch has not matched the prolonged attention of the Johnson administration that overcame this bureaucratic resistance. Finally, societal conceptions of war, instilled during World War II and reinforced by the purported failures in Korea and Vietnam as well as the exaggerated success in the Gulf War, tend to reserve the battlefield for the warrior alone, free from political interference and noncombatant complexities at the tactical level, and supported by the nation's full might. Insurgency violates this model: it is an intimately political form of warfare in which fighter and bystander are interspersed, with limits on use of force. Moreover, insurgency must be met by American civilians as well as Soldiers.

Since American society and its leaders have been slow to accept COIN as war, the government has not applied the joint model of unity of command to our multiagency efforts in Afghanistan, instead accepting a weak surrogate, "unity of effort." In the place of the imperative language of unified authority, the doctrine and directives for the disparate departments urge cooperation, coordination, and consensus, the soft tools of combined warfare. These mechanisms are the strongest available to manage an unwieldy coalition of sovereign state entities, such as the International Security Assistance Force (ISAF),[3] but are far from the strongest available for our own COIN efforts, which still account for half of the international involvement in Afghanistan.

While collocation has brought great benefits to PRTs, the lack of unity of command prevents further integration of the teams. Unified authority would eliminate the long, multiple chains of remote management which impede decision. Additionally, a clear command structure would reduce the role of personality which now unduly influences leadership dynamics among the three senior PRT officials.

Most importantly, unity of command would couple responsibility and authority. Today's model of tripartite command gives each representative the authority to act in his own sphere: the USAID representative runs reconstruction, the State representative directs political programs, and the military team leader is responsible for security. But in COIN, as the widespread use of the Commander's Emergency Response Program (CERP) demonstrates,[4] action in one sphere affects all three. Each representative, reporting to a distant senior, acts to improve his department's sphere, with less concern about the significant effects of that activity in the other two. By uniting command at the provincial level, a single PRT leader, with authority to direct action in every sphere and responsible for effects province-wide, could appropriately manage the broad impact of each decision beyond its bureaucratic sphere of origin.

In Afghanistan, the United States should build on the CORDS concept, uniting all our civilian and military efforts. The American command should designate one team leader of each PRT. In stable provinces, a civilian should lead, with a uniformed deputy; where significant combat continues, a military officer should lead, assisted by a civilian deputy. The Regional Commands should adopt a similar model. At the national level, a civilian ambassador, aided by a general as a deputy and an ample staff, should direct all American activities in Afghanistan through the Regional Commands and PRTs.

This monograph will focus on American efforts in the country, rather than those of the broader coalition. If the United States can take complete responsibility for two of ISAF's four regions and resurrect the solution it devised in Vietnam, it may influence by example the command structure of the coalition.

Diffuse command is not the only challenge we face in Afghanistan. In theater, the greatest obstacles are the tenacity of the enemy and the low capacity of the Afghan government; other impediments include coalition dynamics, organizations ill-suited for COIN, and a lack of familiarity with the host nation. Furthermore, departmental divisions in Washington, exacerbated by congressional loyalties, impede our multiagency effort. Nor is the lack of unified authority the only problem with PRTs; with so few teams and so few civilians, progress will remain slow. I limit my scope to ambiguous management because this organizational problem requires almost no new resources to remedy, its correction will quickly address oft-cited problems, and the solution simply makes sense.

ENDNOTES

1. This monograph uses "agency" and "department" interchangeably to refer to the Defense Department, Department of Agriculture, State Department, and USAID. USAID is technically part of the State Department but largely functions as a separate department in Afghanistan.

2. Such as USAID programs run from Saigon.

3. ISAF is the international coalition leading security efforts in Afghanistan.

4. CERP authorizes military commanders to fund humanitarian and civic projects to alleviate suffering among locals and decrease the likelihood of continued violence. This is traditionally a role that might fall to a development organization such as USAID.

SECTION I. THE STAGE

For 2 1/2 centuries, the government in Kabul has struggled to rule its realm. Regional and global powers have interfered with the country's affairs, pitting Afghan against Afghan by exploiting ethnic tensions along borders dictated by colonial interests. Today's insurgency once again undermines the state, hindered and aided by external powers, as it seeks to extend its writ to the hinterlands. Violence has increased as the population, disappointed with a young government's inability to provide security, lacks the confidence to reject the guerrillas.

After quickly sweeping the Taliban from power and al Qaeda from its sanctuaries, American forces and coalition partners have extended their presence beyond the capital and into the provinces. In the south and east,[1] however, insurgents have accelerated their campaign to undermine the central government, with much of the rise in violence occurring since the beginning of 2006.[2] Suicide attacks, seldom seen in Afghanistan before the American invasion, are now frequent.[3]

THE STATE AND THE INSURGENCY—THEORY IN THE AMERICAN CONTEXT

Although unprepared for this form of warfare that dominates the Global War on Terror (GWOT), American forces were involved in counterinsurgencies throughout the 20th century.[4] In these efforts, the U.S. Government tried to support nascent or recuperating states against an insurgency, a "protracted struggle conducted methodically, step by step, in order to attain specific intermediate objectives leading finally to the overthrow of the existing order."[5] The local government, as assisted by its American patron, competes against the insurgency for legitimacy in the eyes of the population. Today, as much as ever, asymmetry characterizes these wars, as rebels employ lightly armed, loosely organized groups of fighters. The supported government benefits from the resources, advice, and sometimes armor and airpower of its superpower sponsor.

Different Models.

Because of their disparities in power, the two sides typically use different strategic models and perceive different vulnerabilities. U.S. culture tends toward a conventional mindset, in which generals try to defeat the enemy by crushing his forces or capturing his capital.[6] Even when combating an insurgency, this mentality is difficult to escape, and enemy body count appeals to many officials as an indicator of success. Insurgents, fighting in their own territory, cannot hope to destroy an occupying American force; rather, they hope to outlast it by eroding American popular support for the expedition. By creating insecurity and attacking development, they seek to undermine the confidence of the local population in its government and fuel resentment of the occupying power. The insurgents expect the superpower's patience to expire as progress stalls and casualties mount. American departure will leave the struggling state vulnerable to overthrow or disintegration.

Each side follows different operational guidelines—American forces, developed for conventional war, have trained and organized for conflicts of maneuver and attrition, in which they sweep the enemy's divisions from the field; they may train and equip their native allies in this image.[7] Acknowledging the might of American forces, insurgents try to avoid pitched battles that would expose them to overwhelming firepower. They hide among the population and harass the occupier and its government ally with small attacks. Through these surreptitious strikes the guerrillas hope to provoke an indiscriminate allied response which will alienate the population from the state and its partner. From the population's acquiescence and intimidation, the insurgent gains shelter, food, and arms, while also depriving the government of the intelligence necessary to distinguish fighter from citizen.

Public Support—The Center of Gravity.

Insurgents are correct to perceive public opinion as the center of gravity. Insurgents attack popular support at three levels: local support for the state (and, by extension, its occupying sponsor) in theater, the home front of the occupying power (I will refer to this American support as "domestic"), and international sentiment. By targeting American domestic sentiment and native opinion, the insurgents hope to force an American withdrawal, leaving the nascent government vulnerable. The primary determinants of American public support for a war include the perceived probability of success, costs (human and financial), and stakes of the conflict.[8] Should success seem too remote, the costs too high, or the stakes insignificant, popular backing will falter.

If the insurgents gain native support, or at least acquiescence, the occupier will be unable to remove the rebel from his popular base. With no local acceptance of the occupation, success is unlikely and the American public will eventually turn against the war. If the insurgents undermine the morale of the American home front by increasing perceived costs, American forces will probably withdraw as officials respond to electoral pressures.[9] Finally, the stakes always seem lower for the occupying power than for the insurgents who portray their struggle as one of national survival, and as the war drags on, this apparent disparity in stakes grows and Americans begin to question the benefit of perseverance.

The insurgency will exploit all these vulnerabilities simultaneously. However, its direct influence is strongest over native support, so it focuses on undermining the popular confidence in the local government and occupation and deterring cooperation through intimidation. The dynamics of native support for the state, without which American public backing will eventually crumble, create an asymmetry of contending determinants which favors the insurgency. To gain local acceptance, American forces must protect a vast majority of the population as the state is not yet capable of doing so; to undermine confidence in the state and its superpower ally, the insurgency need not capture any territory or destroy conventional units, but rather only attack a few citizens periodically to spread insecurity among all. To win local confidence, U.S. forces must reconstruct the country; the insurgency only need sabotage projects to puncture the promises of the occupation. In seeking to establish the legitimacy of the new government, the occupation may organize elections, and the new officials must endure voters' inevitable disappointments regarding the pace of reconstruction; the insurgents only need to intimidate candidates and disrupt voting to diminish the faith of the public in the new system.

The American Response.

Despite its conventional dominance, economic power, and technological advantages, the United States faces significant disadvantages in counterinsurgency (COIN). Our conventional concept of war as the destruction of an enemy army loses relevance where no conventional force exists. Instead, the American government, like the rebels, must make war indirectly by fostering native support. This approach has the best chance of prevailing in the field, and for justifying the war at home.

To gain local support, or at least acceptance, the American government must pursue a COIN strategy by combining military, political, and economic efforts at the community level to separate the insurgency from its popular base. Successful COIN campaigns rely on tactical flexibility instead of simple preponderance of force; the employment of the minimum force necessary to avoid alienating the population; and the topic of this monograph, civil-military integration, which ensures that all elements of the occupation's power are focused on the same policy without undermining one another.[10]

COIN Operations.

Adapting to their environment, military formations must disperse in the villages to provide security and gain knowledge of social dynamics as well as military intelligence. Every military action has an immediate political ramification at an intimate level; no longer do the two realms reside in separate spheres as depicted in the orthodox American model of warfighting. The aggressive driving of a squad on patrol, which reduces the risk of immediate attack by a suicide bomber, may also alienate a community living near the base. In the future, rather than warning soldiers of recently planted improvised explosive devices (IEDs), villagers may remain silent.

Residents expect not only a judicious use of force from the occupation, but also a significant improvement in their pre-war living conditions. This burden of expectations is especially heavy for a superpower—if the strongest country can quickly vanquish a tyrant, why does it fail to turn on the lights? In addition to extending security, often in the same areas where it is fighting insurgents, the occupation must fulfill basic humanitarian needs, deliver services, rehabilitate infrastructure, and instill these capabilities in the redeveloped state. While specialized military personnel such as Civil Affairs, Military Police, Special Forces, and the Corps of Engineers can address these needs, the American government has civilian agencies (Department of State, Department of Agriculture [USDA], U.S. Agency for International Development [USAID], etc.) whose employees have greater expertise in these forms of assistance.

As each American agency pursues its share of COIN, departmental personnel may begin to see their particular mission in isolation from the overall war effort. USAID contractors might focus on a needs-based disbursement of assistance, regardless of the loyalties of the recipient communities. Military units might focus on capturing a particular cell of insurgents deeply embedded in a community, ignoring the implications for delicate political negotiations within the district government. To avoid these tendencies, the various agencies of the U.S. Government must not only coordinate, but must integrate, their programs so that each complements, rather than undermines, the other. To facilitate this integration, the most successful COIN efforts have had a unified civil-military chain of command under a single manager.

TODAY'S CHALLENGES IN AFGHANISTAN

The Afghan insurgents do not form a coherent movement, but rather a mix of Islamist remnants, warlords, drug runners, and bandits. Motives for violence vary—some who seek political power may be co-opted; others simply profit from disorder and illicit economies and will resist the extension of state authority; the most radical would redefine justice and usurp its enforcement. But all seek to diminish popular confidence in the government of President Hamid Karzai and enhance their own political power. These varied interests play out through Afghanistan's ethnic heterogeneity and tribal culture, which impede the development of national unity.

To counter these forces, the occupation must partner with the underdeveloped forces and ministries of the Afghan government. The occupation leadership must carefully calibrate its transfer of power to the fledgling Afghan government, which may enhance the legitimacy of the campaign but also decrease the delivery of services to the populace, as low capacity, corruption, and inefficiency hamper the new institutions.

These conditions would challenge any occupation; the obstacles of combined warfare complicate efforts in Afghanistan. As American troops are stretched by Operation IRAQI FREEDOM, the North Atlantic Treaty Organization (NATO) provides a growing portion of the International Security Assistance Force (ISAF) in Afghanistan. Each member brings its own national stipulations,[11] as well as varied logistic capabilities. These differing responses create confusion for Afghan partners, creating seams which insurgents may exploit. Coalition difficulties extend to the partnership with Afghan forces. Since local forces continue to lag their coalition counterparts in equipment, capability, and professionalism, the coalition must seriously consider whether to shift the brunt of security efforts to Afghan units, which might not be as effective, or continue to lead the fight at the risk of creating dependency.

Conventional Bias.

In addition to the particular obstacles posed by Afghanistan and the challenges of coalition warfare, characteristics of American society, government, and military have been part of the problem. Recalling the success of the Gulf War and confident that American conventional superiority had increased in the decade following it, the U.S. Government and public felt that the military, empowered by a revolution in military affairs (RMA), was suited for wars large and small. However, the RMA military was the product of late-Cold War strategy and doctrine for the defense of Europe, designed to wreak conventional destruction of atomic proportions on a Soviet opponent. In the opening stages of Operation ENDURING FREEDOM, the technological advantages of stealth, precision, and battlespace awareness seemed equally applicable to a small war. In little more than a month, the American-led coalition expelled the Taliban and its al Qaeda confederates from Afghanistan's major cities. Violence remained at low levels for the next 18 months, and just as U.S. forces crushed their Iraqi counterparts and captured Baghdad in 3 weeks, the RMA military again demonstrated its power across the combat spectrum. The U.S. Government continued to indulge its binary conception of war and peace—as the military had fought and defeated the enemy, it was now the duty of the civilian government, international organizations, and nongovernmental organizations (NGOs) to reconstruct the country.

With this confidence, the government and the military did little to prepare for "insurgency"—through 2003, the military in Afghanistan did not use the term in theater, preferring "counterterrorism" for any fights against violent factions.[12] The distinction is important—generals focused on hunting extremists, rather than securing the population. As violence has increased in Afghanistan and Iraq, the military has begun to recognize the nature of the threat, and altered its training and doctrine to prepare its soldiers for COIN. In addition to chasing terrorists and training native forces, the military has become increasingly involved in reconstruction and securing the population.

Although the military had expected to cede reconstruction to civilian agencies, it has assumed much of this burden out of necessity. The State Department, USDA, USAID, Department of Justice, and other civilian agencies lack the capability to deploy quickly. Nor are these departments designed to support a long-term theater presence—after September 11, 2001 (9/11), Congress did not significantly increase their manning. As peacetime national security structures proved inadequate for insurgency, an over-militarized response ensued.

Provincial Reconstruction Teams.

In part to address theses problems, the U.S. Government has expanded the civil-military Provincial Reconstruction Teams (PRTs) since their creation in 2002. These teams had a three-fold mission: (1) to establish security; (2) to extend the writ of the Kabul government to the provinces; and (3) to facilitate reconstruction. PRTs collocated soldiers and civilian personnel on one base to coordinate distinct departmental programs. By the fall of 2003, four teams were in the field with eight more in 2004; and 25 today. A dozen are American, with just under 100 personnel each; manning on coalition teams varies.[13]

Nearly everyone on a PRT is uniformed; civilians include one representative from the State Department, USAID, and sometimes the USDA. There are also a few Afghan interpreters and an Afghan Ministry of Interior official. By consolidating these personnel in one location, the United States has tried to integrate the diplomatic, economic, and security responses to the insurgency.

Despite the rising violence, PRTs have contributed significantly to Afghanistan's progress.[14] However, three obstacles have hampered the effectiveness of the teams. First, civilian participation is low—the team structure calls for a single representative from each of three departments, and not every PRT is even able to fill these three slots. Second, ISAF has too few teams—25 are inadequate for a country of 600,000 square miles and 30 million people. Finally, the lack of an integrated military-civil chain of command (even on the American PRTs, which are free of the demands of coalition warfare) diminishes the coherence of the American COIN response. This monograph examines diffuse management in Afghanistan, the American response to a similar problem in Vietnam, the effects of ambiguous command, and the benefits of unified authority.

ENDNOTES - SECTION I

1. In 2003, the UN Department of Safety and Security assessed a few pockets of the region as "high risk." By 2005, these had proliferated considerably, and in some cases ratings deteriorated to "extreme risk." In the following year, more districts destabilized, and by June 2006, nearly one-third of the country was described as either high or extreme risk. The trend continued in 2007. See "Countering Afghanistan's Insurgency: No Quick Fixes," *International Crisis Group – Asia Report No. 123,* November 2, 2006, p. 28.

2. In 2006, the last year with complete data, there were 1,677 attacks with IEDs (up from 783 in 2005), 4,542 armed attacks (compared to 1,558 the previous year), and 4,000 deaths, the highest total since 2001. See Seth Jones, "Pakistan's Dangerous Game," *Survival,* Vol. 49, No. 1, 2007, p. 24. The rate of attacks on schools doubled, from 110 in the first 6 months of 2006 to over 400 in the 11 months following. See Joanna Wright, "Taliban Insurgency Shows Signs of Enduring Strength," *Jane's Intelligence Review,* Vol. 18, No. 10, 2006, pp. 24-31; and "Slide Show: War Enters the Classroom," *New York Times,* available from *www.nytimes.com/pages/world/index.html.*

3. In 2002, there was one suicide attack; 2 in 2003; 6 in 2004; and 27 the next year. In 2006, 139 suicide bombers struck, exceeding the historical total. See Jones, "Pakistan's Dangerous Game," p. 24.

4. U.S. forces have fought or assisted local governments against insurgents in the Philippines (three times), Latin America (with the Marine Corps in the interwar period), Greece, Vietnam, Colombia, El Salvador, and Somalia. See Kalev Sepp, "Best Practices in Counterinsurgency," *Military Review,* May-June, 2005, p. 8.

5. David Galula, *Counterinsurgency Warfare: Theory and Practice*, New York: Praeger, 1964, p. 4.

6. The euphoria following the fall of Baghdad in April 2003 reveals the primacy of this American conception of war.

7. For an incisive analysis of this phenomenon during our last major COIN, see Andrew Krepinevich, *The Army and Vietnam*, Baltimore, MD: Johns Hopkins University Press, March 1988.

8. Eric V. Larson and Bagdan Savych, *American Public Support for US Military Operations from Mogadishu to Baghdad*, Santa Monica, CA: RAND, 2005, p. 14. For more analysis on American public support for war, consult Peter D. Feaver and Christopher Gelpi, *Choosing Your Battles: American Civil-Military Relations and the Use of Force*, Princeton, NJ: Princeton University Press, 2004; and John Mueller, *War, Presidents, and Public Opinion*, New York: John Wiley & Sons, Inc., 1973.

9. Paradoxically, the first two factors in public support may work against one another. By minimizing perceived costs, American forces may decrease their chances of success. Consider a risk-averse strategy which harbors soldiers on large bases, thus decreasing casualties, while disorder spreads "outside the wire," causing the people to lose confidence in the government and occupation.

10. For more on these components, see John Nagl, *Learning to Eat Soup with a Knife: Counterinsurgency Lessons from Malaya and Vietnam*, Abington, UK: Praeger Publishers, October 2002.

11. Coalition members have different rules of engagement and combat radii.

12. David Barno in discussion with author, July 18, 2007.

13. Canada, UK, Netherlands, Italy, Spain, Sweden, Germany, Norway, Lithuania, Hungary, and New Zealand all lead PRTs.

14. PRTs have contributed to fair elections, a dramatic increase in education, and improved essential services.

SECTION II. AFGHANISTAN BY DESIGN

ORIGINS

On December 20, 2001, the United Nations (UN) Security Council passed Resolution 1386 establishing ISAF, which the UN tasked with securing Kabul, while American-led forces conducted counterterrorism operations in other parts of the country. President Karzai was pleased with the progress of ISAF in the capital and hoped to extend its civilian-focused peace-keeping to the provinces. Since ISAF nations were initially unwilling to provide troops beyond Kabul, the United States developed the PRTs to stabilize the provinces.[1] The first PRT became operational in Gardez in November 2002. Within a few months, teams deployed to Bamian, Konduz, Mazar-i-sharif, Kandahar, and Herat. The new PRTs expanded the work of the Army's Coalition Humanitarian Liaison Cells and Civil Affairs Teams, and sought to increase participation of civilian agencies in the new stabilization program. However, an unclear mission, low civilian involvement, and limited resources hampered the first PRTs.

To strengthen the program, the American Embassy promulgated a three-part mission for PRTs in February 2003. The teams were to help establish security, extend the authority of the Kabul government to the provinces, and assist reconstruction. Through these goals, the teams would expand the capacity of the Afghan government while maintaining a low foreign profile. The Embassy formed a PRT Executive Steering Committee comprising the heads or senior deputies of the Ministry of Interior, other Afghan ministries, UN Assistance Mission to Afghanistan (UNAMA),

Combined Forces Command-Afghanistan (CFC-A),[2] and ISAF; however, the Steering Committee did not have real executive power, as authority over each team remained with the sponsoring government (still predominantly American during the program's early stages).

In June 2003, the National Security Council Deputies' Committee approved the expanded program, as well as the proposed division of labor for the three primary participating agencies. The Department of Defense (DoD) would improve security in the area of responsibility (AOR), provide all logistical support, and provide force protection for all PRT members, including civilians. The State Department representative was responsible for political outreach and reporting. A USAID official took the lead on construction. Together, the three representatives on each PRT would approve all reconstruction projects and coordinate them with local Afghan officials.[3]

Although coalition officials have not established formal doctrine for the program, several documents have augmented the guidance of the 2003 Deputies Committee memo. The PRT Executive Steering Committee Charter (December 2004), the PRT Terms of Reference (January 2005), and most recently the PRT Handbook (with a third edition released earlier this year), have described guidelines for the teams. The PRT Handbook, jointly signed by the Commander of ISAF, the Senior Civilian Representative of NATO, and the UN Special Representative, maintains the three original goals of the PRT program, noting that the PRTs have a special obligation to lead reconstruction in areas too dangerous for traditional aid organizations. It also asserts that the Executive Steering Committee should provide strategic guidance, and describes the new command relationship of PRTs within ISAF—for military matters, they report to and receive support from the four Regional Commands (RCs). The capital of each team's sponsoring nation, however, still dictates rules of engagement, sets force employment restrictions, and directs reconstruction programs.

THE CONCEPT

Structure.

The personnel and command structure of each team vary between nations. The U.S. model comprises a core of approximately 80 Americans and a few Afghans. The PRT Commander, until recently, was a Civil Affairs officer.[4] With the help of his staff, he leads the military contingent, advises local Afghan officials, and hosts planning meetings with the regional UNAMA office as well as NGOs. Each PRT has two Civil Affairs sections, of four soldiers each. One group makes frequent trips into the province to assess reconstruction projects; the second runs a Civil-Military Operations Center, which coordinates programs with other aid actors in the province. A Police Training and Assessment Team (three MPs) assists local police forces. The military contingent also has a psychological operations (psyops) unit, explosive ordnance disposal team, intelligence cell, medics, aircraft support personnel, and an administrative and support staff of about 20 soldiers. A platoon of 40 soldiers provides security for the compound and trips into the countryside.

As envisioned, an American PRT should also have a State Department Foreign Service Officer (FSO), a USAID representative, and a USDA expert. The FSO serves as the political advisor to the PRT Commander, the Governor of the province, and other local officials. Deployed FSOs also report provincial political dynamics to the Embassy. The USAID representative works closely with the Civil Affairs teams, advises on all development work, and coordinates projects with the local government and area NGOs. The Agriculture employee provides veterinary and horticultural assistance to locals, a critical service given that 80 percent of the population is engaged in farming.

An officer from the Afghan Ministry of Interior (MOI) advises the PRT on local politics and helps intelligence efforts. Most PRT commanders found the MOI's assistance indispensable.[5] Three or four Afghans serve as interpreters.

Chain of Command.

The PRT Commander is responsible for the care (food, housing) and logistical support of all team members. The commander has been until recently a Civil Affairs officer and thus been assigned command of only the Civil Affairs team members (who, like the commander, are often reservists). Other military elements (force protection, intelligence, psychological operations, air detachment, etc.) are attached to the PRT to support its efforts—while they may receive broad mission guidance from their own chain, the commander has varying degrees of tasking authority over these units. While he does not have Uniform Code of Military Justice (UCMJ) authority and does not write evaluations for these personnel, he could likely remove a miscreant through his influence with the component chain of command.[6]

Funding, evaluation, assignment, and programmatic authorities for civilians rest with their parent agency in Kabul (although the PRT representatives retain some autonomy because of their isolation). USAID and State have different reporting systems. Deployed FSOs report directly to the State PRT Office in Kabul. USAID, however, has mirrored the changes in the military command structure, adding Development Advisors at the Regional Command (RC), who supervise programs at the PRT and report to USAID's Civil-Military Affairs Office in Kabul.

Collocation is the greatest benefit of the PRT program. Representatives of USAID, State Department, USDA, and the military bring unique expertise to the team and, as they learn each other's perspective, can jointly develop a program geared towards the three components of the PRT mission and tailored to local conditions. A cramped, isolated base fosters a cooperative spirit, and the focus on a single province and its particular problems creates a common outlook among representatives now removed from their parent department. The constant intercourse between the representatives who live, eat, and socialize together introduces each to the other's agency, and allows access to each department's resources.

Multiple reporting routes bring some advantages to the teams. As each department in Kabul has its own chain to the PRTs, the director and mission staff receive timely accounts of the local effects of provincial and national programs. Reciprocally, team members benefit from their colleagues' quick access to their Kabul headquarters. One RC Commander noted that "if you have the State Department guys tied to State's Headquarters, you get a good link to that higher authority and that expertise."[7] Some respondents cite the ability to leverage these separate chains of command; if one representative is having trouble accessing departmental funds, a colleague from a different department might send word up a second chain of command, and a meeting by senior officials in Kabul could free up money.[8]

The teams rely on several sources of funds to finance reconstruction projects. The military has access to Commanders Emergency Response Program (CERP)[9] and Overseas Humanitarian Disaster Assistance and Civil Aid (OHDACA). USAID has contributed Quick Impact Project (QIP) funds, which focus on small infrastructure improvements, and has recently transitioned to a new type of funding, the Local Governance and Community Development program, designed to increase the capacity of the local government. The State Department has little money earmarked directly for projects, since its work on political development is not as expensive as reconstruction.

To coordinate the programs of different departments, a PRT Commander will often develop a system of regular meetings with other team members as well as provincial actors. One PRT leader convened a nightly internal staff meeting with all civilians and senior personnel of the various military attachments to review the day's activities and plan for the next day. This commander also held a weekly Operations Synchronization meeting which included the PRT staff as well as NGO and UN officials.[10] Through these meetings, smoothly functioning teams coordinate different funding sources to provide continuity to provincial programs. An officer with PRT experience noted that teams might start construction of a school with readily accessible CERP accounts, and then, once approved, draw on USAID's Quick Impact Project funds.[11]

By using these internal systems which accommodate multiagency representation, most PRTs develop a consensus, rather than command, decision process. Participants note that this cooperative management style, when functioning well, addresses concerns of each department. One deployed FSO recounted a visiting ambassador asking who was in charge. The team leader responded "There are four of us," including the FSO, the USAID representative, and the CO of the attached infantry battalion. The State representative praised the leadership climate—"it got to the point where we were practically of one mind."[12]

NATION TEAM

A Shift in Strategy.

As PRTs expanded in late 2003 and increased coordination between civilian agencies and the military at a tactical level, a new commander of American forces tried to integrate civil-military operational efforts at the national level. When then-Major General David Barno, USA, arrived in October 2003 to lead the CFC-A, comprising nearly all American forces in the country, he declined to craft a narrow military mission statement. Rather, he directed his forces to adopt a broad mission: make U.S. policy goals—the creation of a stable, democratic, unified state—succeed in Afghanistan.[13]

To accomplish this new mission, he changed the military's operational focus. Uttering a word that most leaders in theater had eschewed, Barno admitted that an "insurgency" threatened the prospects of American success in Afghanistan.[14] To address this danger, he directed American forces to shift their efforts from counterterrorism to COIN. Although counterterrorism remained important, it would assume a supporting role. Following a classic COIN strategy, Barno designated the population and its security, instead of violent Islamists, as the center of gravity. The new CFC-A Commander embraced nation-building, a mission that the previous command had explicitly declined. He considered PRTs, with their mission of security, governance, and reconstruction, an apt instrument for the new population-centric strategy and added resources to the teams.

The Move to Kabul.

When he took command of forces, Barno felt that tactical considerations were subsuming operational goals. To implement his new strategy, he wanted "to establish Unity of Command and Unity of Effort, of which we had none."[15] Barno moved his military headquarters to Kabul in order to create this coherence. Previously, the senior American general resided on a large combat base at Bagram, while civilian leaders (Afghan, American, and those of other nations), UN heads, NGO directors, and the ISAF contingent quartered in Kabul, about 35 miles away. General Barno felt that the geographic separation impeded integration of the military operational planning with broader American strategy and Afghan government concerns. Discussing the potential move, a senior commander warned that "Kabul will consume you"[16]—exactly Barno's intent.

While the top American military and civilian leaders would never achieve unity of command in Afghanistan, the general's move did facilitate coordination between the Embassy and his staff. Barno and Zalmay Khalilzad quickly forged a strong relationship. The general found that he shared a "common view of the fight" with the new Ambassador, who endorsed the new COIN strategy.[17] Acknowledging the primacy of the political in counterinsurgency, Barno saw himself as the supporting element on behalf of Khalilzad's mission. To cement this new partnership and bring together their respective staffs, the general moved his headquarters into the Embassy; quarters elsewhere in Kabul, while an improvement on Bagram, were insufficient. This collocation was of "huge importance"[18]—easy access no matter the hour and shared workdays, meals, and recreation, helped to integrate the policies of the two leaders and their staffs. This common quartering also signaled to the Afghan government, subordinates in the field (civilian and military), and the Washington departments that the Embassy and military were pursuing a unified strategy. The shared Embassy also allowed Barno to second military planners to the Ambassador's chronically undermanned staff.[19]

Relocating to Kabul helped CFC-A's international outreach as well. As he formed a detailed document describing his new COIN strategy, General Barno circulated drafts throughout the international leadership in Kabul, gathering input from all USG bodies, UNAMA, and the Afghan government.[20] Building this consensus partially compensated for the lack of an international combined command.

Reorganizing the Military.

To clarify the responsibility of military units, General Barno altered the structure of his field commands. Previously, Combined Joint Task Force 180 (CJTF-180) in Bagram had directed all forces nationwide. Maneuver battalions reported directly to Bagram, with no fixed area of responsibility. Other military units worked primarily within their functional chains of command (e.g., air, logisitics), reporting through these to Bagram. PRTs reported to the Coalition Joint Civil-Military Operations Task Force (CJCMOTF) in Kabul, which in turn answered to Bagram.

Barno split the country into four regions, assigning responsibility for all military units within each region to Regional Commands (RC). His goal was "to get coherence, ownership of results, by assigning geographic battlespace to battalions and brigades so that now units would report to a specific person [the RC Commander]."[21] Barno transferred operational authority over the PRTs' military elements[22] from CJ-CMOTF to the RC. He also gutted CJCMOTF over the objection of many of its Civil Affairs staffers, sending many personnel to the eight new PRTs deployed in the first half of 2004.[23] The consolidated authority of the RCs increased the coordination between the sometimes conflicting operations of maneuver battalions, which focused on strike missions, and PRTs, which focused on stabilization. One FSO praised the new structure, noting that "the PRT commander reports back to this chain that is now integrated with the warfighting element and the reconstruction effort element and that's good, because it . . . seemed to reduce the frustrations that poor PRT commander is fighting."[24]

The new chain of command effectively separated operational and administrative control of the PRTs' military components between the RCs and CJCMOTF, respectively. One RC Commander surmised that sensitivity to bureaucratic culture motivated the split reporting chains—PRT leaders were more comfortable being rated by a Civil Affairs officer (rather than a combat arms commander at the RC), most of whom were attached to CJCMOTF.[25]

ENDNOTES - SECTION II

1. Michael J. McNerney, "Stabilization and Reconstruction in Afghanistan: Are PRTs a Model or a Muddle?" *Parameters*, Winter 05-06, p. 32.

2. From 2003 through 2006, CFC-A was in charge of all American forces in Afghanistan, with the exception of some special operations forces.

3. Paul Ciminelli *et al.*, "Provincial Reconstruction Teams in Afghanistan: An Interagency Assessment," April 5, 2006, pp. 9-10.

4. Over the past year, the Navy and Air Force have begun to provide most PRT Commanders.

5. Robert Perito "United States Institute of Peace Special Report 152: The U.S. Experience with Provincial Reconstruction Teams in Afghanistan—Lessons Identified," Washington, DC: United States Institute of Peace, October, 2005, p. 6.

6. PRT Commander in discussion with author, July 12, 2007.

7. RC Commander in discussion with author, July 16, 2007.

8. PRT Representative, interview by Peter Bolton, July 29, 2005, interview 36, transcript, United States Institute of Peace and Association for Diplomatic Studies and Training.

9. CERP gives a military commander authority over a defined amount of money for reconstruction and humanitarian response.

10. PRT Commander in discussion with author, July 12, 2007.

11. PRT military officer in discussion with author, July 16, 2007.

12. PRT Representative, interview by Barbara Nielsen, August 31, 2005, interview 46, transcript, United States Institute of Peace and Association for Diplomatic Studies and Training.

13. Barno commented that "this is a philosophy issue. Many military leaders think that you should put the military mission in a box, that's all they're concerned about." David Barno in discussion with author, July 18, 2007.

14. David Barno in discussion with author, July 18, 2007.

15. *Ibid*.

16. David Barno Draft Working Paper, p. 6.

17. David Barno in discussion with author, July 18, 2007.

18. *Ibid*.

19. A great benefit from joining the civilian and military staffs was the creation of a working group that was solely responsible for tracking all American government funding of efforts in Afghanistan. Barno staffer in discussion with author, July 26, 2007.

20. David Barno in discussion with author, July 18, 2007.

21. *Ibid*.

22. State, USAID, and Agriculture personnel still reported to their individual directors, not the RC.

23. CJTF-180/76 staff in discussion with author, October 14, 2007.

24. PRT Representative, interview by Peter Bolton, July 29, 2005, interview 36, transcript, United States Institute of Peace and Association for Diplomatic Studies and Training.

25. RC Commander in discussion with author, July 16, 2007.

SECTION III. CIVIL-MILITARY REALITY

The close coordination of the Embassy and CFC-A and the rapid expansion of PRTs from 2003 to 2005 improved cooperation between the civilians and the military. However, some hopes of that period have receded, due in part to the expected complications as the coalition grows; of greater concern, fundamental problems in the chain of command, which have never been addressed, persist.

PROBLEMS AT THE NATIONAL AND REGIONAL LEVELS

Changes in Kabul.

As coalition participation has increased, ISAF has expanded its area of responsibility. In 2006, the American government disbanded CFC-A and ceded responsibility for all stability and security operations to ISAF, under NATO command. The alliance includes 37,800 troops deployed in 4 regions (RC East, South, North, and West) and a capital area.[1] Each region has a 1 or 2-star officer who reports to the ISAF commander, who at this writing was General Dan McNeill, USA. Roughly 15,000 of ISAF's soldiers are American, falling under RC East commanded by Major General David Rodriguez. These include 1,000 PRT personnel.

An additional 11,000 American Soldiers comprise OEF[2] efforts in Afghanistan. These troops fall under 3 commands: CSTC-A[3] is responsible for training and equipping Afghan security forces; the Joint Special Operations Task Force leads counterterrorism efforts; and the National Support Element provides logistics and administrative support for American troops in theater. Each of these commands reports to CENTCOM, although SOCOM shares responsibility for JSOTF.

At the national level, the American Embassy is not as closely tied to ISAF as it had been to CFC-A. This change is appropriate; the United States has lost its unilateral sway in Afghanistan, and ISAF should coordinate more broadly. However, the separation between the American command and the Embassy preceded the turnover to ISAF. When Lieutenant General Karl Eikenberry took command of CFC-A, he moved his staff out of the Embassy to a separate compound in Kabul. With the departure, Eikenberry's interactions with the new Ambassador, Ronald Neumann, decreased as did the integration of their staffs. Neumann did not object to the departure of CFC-A Headquarters, and the relationship between the Ambassador and the new commander was not as close as that of their predecessors.[4]

The separation between the Ambassador and the American military commander had significant effects in addition to the immediate deterioration of joint planning. First, coordination looked less important to the Afghan government, as well as to American personnel.[5] Second, contentious issues such as a border security plan and a transition of police training from State Department to DoD, which the Ambassador and CFC-A commander had previously resolved, were now referred to Washington.[6]

The new Ambassador lacked the authority and clout of Khalilzad, who had served not only as Ambassador but also as Special Presidential Envoy to Afghanistan, enjoying frequent access to the U.S. Vice-President and President.[7] However, the fundamental ambiguity of the authorities of the Chief of Mission (CoM) hindered both Ambassadors. State Department, USAID, and military officials in Afghanistan and

Washington have different understandings of the authority of the Chief of Mission.[8] Some contend that the Ambassador's position as Chief of Mission bestows a degree of directive power over other civilian departments. Others claim that he may influence, but not direct, the programs of different departments. One senior State Department official observed that the Ambassador's executive power depends on the individual's personality and the "clout" of the State Department at that time, adding that "even in the best run country team which is fully integrated, there is a certain amount of stovepiping and direct reporting to the parent office in DC."[9]

Even under the Barno-Khalilzad team, the practice of interagency coordination fell short of the vision. Barno praised the integrated national view he received by daily exposure to every department's reporting chain, noting that his picture was much more complete than that available to any other military officer.[10] The field did not enjoy such smooth coordination. Information did trickle down, but often circuitously—an RC Commander noted with no irony that the FSOs on the PRT:

> have a direct link to the ambassador's office which has a direct link back to Washington. They were able to find out at least from my perspective what the U.S. Government's intent was for a specific area, but they were also able to talk to government officials, get information out of Kabul that the ambassador was getting from the president and let us know at least what the central government had planned for an area.[11]

A more efficient chain might transmit the information directly to the RC Commander, instead of requiring him to depend on a subordinate PRT for an understanding of American strategic intent. Worse, whether shared or not at the national level, plans are sometimes not extended to the field through any chains. One PRT commander described an unannounced Department of State anti-drug operation employing Afghan commandos, which upset villagers. The governor demanded, "Can you meet with 300 people that are upset because they had some military come in and do this drug bust and nobody knew about it?" The PRT commander, as uninformed as the provincial officials, could give no detailed explanation.[12]

Regional Dynamic for the PRTs.

Although Barno's system of Regional Commands strengthened battlespace ownership, it has not given the Regional Commanders true authority over their PRTs. National restrictions inhibit cooperation between teams, and diminish the ability of RCs to direct and integrate operations within their regions. Additionally, the RC has formal command only over the military elements on the PRTs (which remain constrained by national caveats); national missions in Kabul direct the reconstruction and diplomatic programs of each team.

Nor has the RC model of area responsibility been applied to the provincial level—American PRTs retain responsibility only for the reconstruction within their province, not for offensive operations executed by maneuver units. So, while the RC largely controls the forces in its region (with the important exception of the embedded civilians on the PRT) in accordance with national restrictions, the team commander does not have control over the province. If the team needs robust assistance for a mission in a more dangerous part of the province, the PRT commander depends on the receptivity of the maneuver element. If he worries that the strike battalion's security sweep might endanger a nascent disarmament deal with a local warlord, he must appeal to the RC. A State Department official who worked at a New Zealand PRT noted the difference—the commander was in charge of the PRT as well as tactical security patrols, "so if something was out of whack, he could realign it."[13] An FSO deployed to an American PRT called for the New Zealand model—"A single integrated command . . . the attempt to distinguish between civil affairs and government support and combat security operations, [being] a false barrier."[14]

Under the current American arrangement, an appeal to the RC often gains little—the PRT commander is countering a military culture that still favors the "kinetic," finding and engaging bands of enemy fighters, over mundane stabilization. One military officer who was stationed at the Embassy and worked closely with the teams described "bureaucratic resistance from some traditional ground combat commander personalities," noting that "I don't think in general the conventional force has done well at learning the lessons of previous deployments as far as reconstruction goes. I think they were much more oriented towards the security piece. . . . At particular points in time, that's a misplaced priority."[15] One PRT official noted "that was the only way we could get that kind of support, if we had a maneuver commander actually calling it in. There was no way a little PRT was going to get that. If you have a maneuver commander who isn't so tuned in . . . who isn't that sensitive, then you're stuck."[16] A team commander, acknowledging command bias, suggested total geographic segregation as a solution—"There was a lot of crossing over, and that was irritating at times, but, hey, there was really no way around it because the command would not change it. I'd put PRTs where you weren't going to have other forces."[17] But this drastic solution would leave the most violent areas without PRTs.

In addition to the cultural tension between tactical and reconstruction missions, PRTs face a deeper institutional bias. Until recently, Civil Affairs (CA) officers, many of whom are reservists, commanded most teams.[18] PRT civilians and RC commanders alike note that CA officers are well-suited for the team's complex mission.[19] However, Soldiers from traditional combat arms, such as Armor and Infantry, sometimes consider Civil Affairs a support unit, its members not quite full warriors. A PRT member described the tension:

> The Army Civil Affairs people sort of bristle when they are treated as people who have to be protected. They will tell you, "We're Special Forces. We are Airborne trained. We're all Soldiers. We're all trained. We're all armed. We protect ourselves, thank you very much. We're not orphan children who have to be shepherded by Army infantry." They take pride in the fact that they are Army and they can protect themselves. Just because they're Civil Affairs doesn't mean that they're not Soldiers. The maneuver companies have a rather different perspective. They tend to think of the PRTs more as sort of civilianized entities and not "real" Soldiers.[20]

A State representative echoed this assessment, commenting that the "PRT commander has to be aware of this—and if he's a CA guy, he'll know this anyway—they can be viewed as the red-headed step children of the military brethren."[21] Even though RC commanders recognize the skill of Civil Affairs personnel in leading

PRTs, they often succumb to the combat arms bias. An FSO noted that:

> The big problem was with the maneuver element. Almost every PRT and maneuver element share an AO. The PRT should be the supported element, but no brigade commander would tell an active Army commander that he supports a Civil Affairs reservist, so he either defaulted to the maneuver element or told the two to work it out.[22]

When the team and maneuver element commanders "work it out," the former will often defer to the latter, who has the coveted position of battalion commander in a military that prioritizes tactical strike. Few maneuver leaders will submit themselves to the tacit control of the PRT.

The combination of cultural bias for kinetic operations and against Civil Affairs units leads maneuver units to dominate teams in some regions. An FSO described the effects of collocation with a maneuver element on one team—"Force protection for the PRT is somewhere down on the list of their priorities and missions and duties and responsibilities, towards the bottom third. So, what were PRTs have now become sort of enhanced civil affairs sections of combat maneuver companies."[23] Another civilian, lamenting that teams "did not have adequate commo-chains back to the actual warfighters," noted that the PRTs often "were on the lowest priority peg for any kind of support."[24] A USAID officer recalled that the PRT Force Protection unit would often be assigned auxiliary security tasks, stripping him of his escort and causing him to cancel missions.[25] During a period when the RC commander tasked the team's force protection component to undertake a tactical mission, a USDA representative complained that the average number of sorties for reconstruction tasks dropped from 4-5 per day to 1-2 per day.[26]

INTRA-PRT DYNAMICS

Lack of Guidance and Uncertain Command.

A 2006 Interagency Assessment charged that the "lack of operating guidance clearly delineating missions, roles, responsibilities, and authorities," causes civil-military tension within the PRTs.[27] For example, the new PRT Handbook is descriptive, not prescriptive; while it outlines organizational relationships with the RC and Executive Steering Committee, the guide notes that internal organization is determined by the sponsoring nation. But the United States still has not formalized the command structure for its team. Members devote much energy to negotiating responsibilities, leaving performance dependent on the particular dynamics of the PRT staff.[28]

Who's in charge? The answer depends on the respondent. Although senior military officers recognize that civilians have separate reporting chains, they are confident that the PRT commander has de facto control over the team, and that this system works harmoniously. An RC Commander noted that:

> civilians worked for the PRT commander just like the military guys did . . . guidance and direction came from the PRT commander. We didn't have guidance and direction from a State guy [in Kabul] to a State guy, from USAID to USAID. Rather, the guy who owned the battlespace, the RC commander, gave marching orders and everyone got on board.[29]

General Barno gave a similar assessment, noting that the "commander of the PRT gave guidance and direction to civilian players, although the civilian players also reported up to the mission . . . I never had DoS or USAID disputes; their reps understood that PRT commanders owned that area and that civilians operated under PRT commander rules."[30]

Most USAID and State representatives recognize the tendency of the military to arrogate power, but maintain that they are under the authority of the military for force protection only. One USAID representative stressed her agency's autonomy, noting that the national director instructed program officers deployed to PRTs to consider themselves an independent USAID Field Office. Describing the declaration in the 2007 edition of the PRT Handbook that all senior military and civilian personnel on PRTs are coequals, she added that civilian PRT representatives received the grade of GS-13/14 primarily to establish their parity with the O-5 military commander. In this egalitarian spirit, she no longer uses the term "PRT commander," but instead refers to the "military lead." However, she admitted that few used her new lexicon.[31]

State Department representatives recall a different theoretical model for the teams, but concede its neglect. A 2004 National Security Council Deputies Committee Memo urged consensus but gave State decisionmaking primacy in all PRT reconstruction issues;[32] however, "neither military nor civilian colleagues put any weight on or even knew of the memo."[33] With reality falling short of the model, State representatives want at least to maintain autonomy. A Foreign Service Officer recounted that when he arrived at the two PRTs where he served, the military commander would say "you're my POLAD [political advisor], you're my

DoS Representative" to which he would demur "I work with you, not for you."[34]

With regard to specific projects, decisionmaking depends on the dynamics of the individual teams. Some veterans describe a leadership council; others portray complete delegation by the PRT commander (interestingly, the concept of delegation implies that the authority originally resides with the colonel); a few recall a military bent for programmatic decisions. With this range of command dynamics, a representative does not know what to expect when he reports to the team, whether he will have to defend his role from encroachment or stir counterparts to action. Within the team, bureaucratic loyalties usually persist—one USAID officer reflected, "It would have been more useful to have a more close-knit organization where you looked at the functions more than you looked at the place the person came from."[35] This parochialism diffuses responsibility, as representatives blame difficulties on inadequate assistance from their colleagues, with the military the most frequent scapegoat.

Although tension often arises over civil-military lines, USAID and State representatives also disagree about programs and resources. The USAID Administrator nominally reports to the Secretary of State, but USAID employees in the field maintain their independence. An RC commander noted that "AID guys are great, but say 'My mission is development. I work for USAID, and my parent HQ is in Kabul.' So there is still stovepiping going on. . . . USAID employees absolutely see themselves as independent of State."[36] One State representative working on a multiagency team in Iraq was shocked to hear a USAID representative declare in a meeting with local officials that he, the USAID individual, was "not a representative of the U.S. Government." The State representative said this is an extreme example of USAID's traditional "We do things on a needs basis" approach that is gradually aligning with national security concerns.[37]

Cult of Personality.

When asked whether the ambiguous command structure is effective, PRT members, operational leaders, and officials in Washington invariably respond that performance depends on personalities. "The bottom line is that everything is ad hoc. If you have the right personalities, it works great; if not, everyone goes separate ways."[38] With no formal interagency integration and thus no on-site command, individual motivation determines PRT performance. An FSO exclaimed, "[H]ere's the thing, the State Department person can just sit there in his office and pretty much collect his pay, like a consumer of food . . . but you also have the capacity to really help your military colleagues a lot."[39] A USAID officer maintained that "people who cared, worked wonderfully; those who just wanted a paycheck didn't care."[40] The Interagency Assessment captured these concerns in its surveys, noting that performance is too dependent on interpersonal dynamics and individuals' appreciation of the mission.[41]

What happens when personalities don't match? "Problems," replies an RC Commander.[42] Complaints from the military and USAID rise up to the RC level; but due to the flat State chain, the issue might simultaneously rise to the mission in Kabul. An FSO tells of a USAID colleague who condemned the PRT mission statement as inadequate and refused to fund any projects. Upon appeal to a supervisor at the State PRT Office in Kabul, "the response was 'no surprise.' I'm not sure if he actually spoke to his USAID counterpart."[43]

The Difficulty of Distant Command.

How was it possible that one of the key members of the PRT team could simply opt out? "It was the same for her, for me, for everyone—if she didn't report something, it was invisible. . . . She created the situation, but she wouldn't have to report it that way."[44] Each departmental mission in Kabul has one formal information link from the province—its representative. Few officers will condemn themselves to their seniors. When a dispute arises, each mission receives an account from its representative; these probably vary, and each department tends to support its deployed personnel.

Continuing, the FSO describes the dynamic in Kabul:

> The senior directors that we have both appealed to are at the Embassy; they see each other all the time, eat together, exercise together. So, if they piss someone off at the Embassy compound, repercussions last for weeks. They don't want to make waves in the capital. If they piss off someone in the PRT [by ignoring the appeal from the field], what is that person going to do? Send an email? Ooooh![45]

The distance of the arbitrating body leads not to resolution, but inaction.

Military Dominance.

The military officer is usually the most influential team member due to preponderance of resources as well as seniority. The PRT commander is responsible for housing, food, and logistical support for all members—control of these resources provides soft influence. Additionally, the team commander approves sorties from the base and decides the priority of trips. Although always striving for consensus, a PRT commander notes that "if we have to cut one move, I make the call. Civilians accepted if a move got cut as well as military guys. Sometimes we have to make a decision and they might not like it, but so be it. It's not based on who is wearing a uniform but rather what was the effect."[46] Approval of moves strengthens the PRT commander's authority in the uneasy balance of power. Finally, the team leader exerts influence through the military's control of the most easily accessible funding, CERP. A State official mused that "AID may have 'the lead' on development, but who's signing the CERP [military authorized funding] check?"[47]

The commander's seniority also inspires deference. Lacking the deployable capacity of the military, the State Department and USAID have trouble filling PRT billets and often rely on younger employees or contractors. PRT commanders are lieutenant colonels or colonels with 2 decades of service. A Defense official observes that "technically, the State person is supposed to be in charge of political affairs. But State can only get a junior FSO out there. A lieutenant colonel won't take orders from a 30-year-old who has never been in the field before. . . . So what tends to happen is the military commander is the commander for the whole PRT."[48]

The 2006 Interagency Assessment describes the military dominance of some teams, especially on those that host maneuver elements, and notes the tendency for some PRT commanders to consider the attached civilians mere advisors. When civilians are excluded from decisionmaking, the mission suffers.[49] One USAID representative, worried that the military mission of the PRT too often took priority over development, saw segregation, rather than integration, as the solution and suggested dividing the security and development responsibility between different PRTs.[50]

PRTs face two sets of obstacles—military bias against the team mission and poor organizational dynamics. A conventional outlook leads sympathetic Regional Commanders to favor maneuver elements over PRTs. Lack of clear command and control makes authority uncertain within the team, performance overly dependent on personality, and supervision too distant to be effective. Combined, these problems lead to an over-militarized response to insurgency. A unified authority would overcome these obstacles, improving our COIN efforts in Afghanistan.

ENDNOTES - SECTION III

1. See map, Appendix 1.

2. Operation ENDURING FREEDOM.

3. Commander, Security Training Command—Afghanistan.

4. Christopher Griffin, "A Working Plan—Hope Isn't the Only Strategy for Afghanistan," *Armed Forces Journal*, available from *www.armedforcesjournal.com/2007/04/2587549*.

5. Barno staffer in discussion with author, July 26, 2007.

6. David Barno in discussion with author, July 18, 2007. Barno Staffer in discussion with author, July 26, 2007.

7. David Barno in discussion with author, July 18, 2007.

8. Multiple interviews in July 2007.

9. Senior State Department official in discussion with author, July 26, 2007.

10. David Barno in discussion with author, July 18, 2007.

11. PRT Representative, interview by Barbara Nielsen, August 11, 2005, interview 43, transcript, United States Institute of Peace and Association for Diplomatic Studies and Training.

12. PRT Representative, interview by Larry Lesser, April 27, 2005, interview 21, transcript, United States Institute of Peace and Association for Diplomatic Studies and Training.

13. FSO discussion with author, July 23, 2007.

14. PRT Representative, interview by Barbara Nielsen, August 31, 2005, interview 46, transcript, United States Institute of Peace and Association for Diplomatic Studies and Training.

15. PRT Associate, interview by Barbara Nielsen, September 27, 2005, interview 52, transcript, United States Institute of Peace and Association for Diplomatic Studies and Training.

16. PRT Representative, interview by Peter Bolton, July 29, 2005, interview 36, transcript, United States Institute of Peace and Association for Diplomatic Studies and Training.

17. PRT Representative, interview by Barbara Nielsen, September 20, 2005, interview 45, transcript, United States Institute of Peace and Association for Diplomatic Studies and Training.

18. With an increased demand in Iraq, Civil Affairs personnel have shifted to a theater where they do not enjoy equivalent command, and Air Force and Navy personnel have filled the gap.

The shift demonstrates the lack of priority the Army, and Civil Affairs more broadly, puts on leading PRTs.

19. RC Commander discussion with author, July 16, 2007.

20. PRT Representative, interview by Barbara Nielsen, July 19, 2005, interview 38, transcript, United States Institute of Peace and Association for Diplomatic Studies and Training.

21. PRT Representative, interview by Peter Bolton, July 29, 2005, interview 36, transcript, United States Institute of Peace and Association for Diplomatic Studies and Training.

22. FSO discussion with author, July 24, 2007.

23. PRT Representative, interview by Barbara Nielsen, July 19, 2005, interview 38, transcript, United States Institute of Peace and Association for Diplomatic Studies and Training.

24. PRT Representative, interview by Peter Bolton, July 29, 2005, interview 36, transcript, United States Institute of Peace and Association for Diplomatic Studies and Training.

25. PRT Representative, interview by Jack Zetkulic, July 20, 2005, interview 32, transcript, United States Institute of Peace and Association for Diplomatic Studies and Training.

26. PRT Representative, interview by Barbara Nielsen, September 30, 2005, interview 49, transcript, United States Institute of Peace and Association for Diplomatic Studies and Training.

27. Paul Ciminelli *et al.*, "Provincial Reconstruction Teams in Afghanistan: An Interagency Assessment," April 5, 2006, p. 13.

28. PRT Representative, interview by Peter Bolton, July 29, 2005, interview 36, transcript, United States Institute of Peace and Association for Diplomatic Studies and Training; and Paul Ciminelli *et al.*, "Provincial Reconstruction Teams in Afghanistan: An Interagency Assessment," April 5, 2006, p. 12.

29. RC Commander in discussion with author, July 16, 2007.

30. David Barno in discussion with author, July 18, 2007.

31. PRT AID Representative discussion with author, July 17, 2007.

32. FSO discussion with author, July 23, 2007; FSO discussion with author, August 1, 2007; and PRT Representative, interview by Barbara Nielsen, August 31, 2005, interview 47, transcript, USIP and ADST.

33. FSO discussion with author, July 23, 2007.

34. PRT FSO discussion with author, August 1, 2007.

35. PRT Representative, interview by Jack Zetkulic, July 20, 2005, interview 32, transcript, United States Institute of Peace and Association for Diplomatic Studies and Training.

36. RC Commander in discussion with author, July 25, 2007.

37. State Department official in discussion with author, July 13, 2007.

38. *Ibid.*

39. PRT Representative, interview by Beverly Zweiben, August 2, 2005, interview 41, transcript, United States Institute of Peace and Association for Diplomatic Studies and Training.

40. AID Representative discussion with author, July 17, 2007.

41. Paul Ciminelli *et al.*, "Provincial Reconstruction Teams in Afghanistan: An Interagency Assessment," April 5, 2006, p. 15.

42. RC Commander in discussion with author, July 16, 2007.

43. FSO in discussion with author, July 23, 2007.

44. *Ibid.*

45. *Ibid.*

46. PRT Commander in discussion with author, July 12, 2007.

47. FSO discussion with author, July 24, 2007.

48. DoD official discussion with author, July 9, 2007.

49. Paul Ciminelli *et al.*, "Provincial Reconstruction Teams in Afghanistan: An Interagency Assessment," April 5, 2006, pp. 12, 14-15.

50. PRT Representative, interview by Jack Zetkulic, July 20, 2005, interview 32, transcript, United States Institute of Peace and Association for Diplomatic Studies and Training, p. 10.

SECTION IV. ANTECEDENTS FROM VIETNAM

Forty years ago, American officials also had difficulty integrating the government's response to insurgency. In 1967, after several years of failed attempts, the Johnson administration implemented the Civil Operations and Revolutionary Development Support (CORDS) program, which formally combined the command chains of every department involved in the COIN effort. This structure created a unified response from the programs of the various depart-ments and substantially improved pacification.[1] By 1971, the insurgency had withered.

A LONG ROAD

Focus on Counterinsurgency.

Early in the Kennedy administration, officials recognized the primacy of COIN. Frustrated with the extremes of inaction versus nuclear strike offered by Eisenhower's doctrine of Massive Retaliation, the new President promulgated Flexible Response, which proposed to meet the nation's threats through graduated force. As wars of national liberation raged, Kennedy instructed the government to prepare for COIN. Encouraging the future Army to embrace these new responsibilities, the President urged the graduating West Point class of 1962 to be ready for a type of war against guerrillas who see "victory by eroding and exhausting the enemy instead of engaging him. . . . It requires in those situations where we must counter it . . . a whole new kind of strategy, a wholly different kind of force, and therefore a new and wholly different kind of military training."[2] He recognized that since the mid-1950s, Army advisers had tried to form the South Vietnamese Army (ARVN) in the image of the American force—a mechanized, conventional army, trained for interstate conflict, not COIN.

Kennedy anticipated bureaucratic resistance, stating, "I know that the Army is not going to develop this counterinsurgency field and do the things that I think must be done unless the Army itself wants to do it."[3] While the Army publicly embraced COIN, it made few changes in its training to incorporate the new doctrine. Realizing the broad demands of this unconventional warfare, the administration also directed civilian agencies to embrace COIN. It initiated an interdepartmental seminar for the Central Intelligence Agency (CIA), State Department, and military, and created a Special Group for Counter Insurgency to draft an interagency approach to COIN. However, the rest of the government was no more inclined to innovate than the military.

Failure to Change.

The President hoped that the adoption of COIN doctrine by the U.S. Government would enhance pacification efforts in Vietnam, and several civilian agencies joined the military in assisting the Saigon government, which was competing with the Viet Cong for legitimacy. Communist insurgents provided health care, schooling, land reform, and infrastructure in the villages they controlled.

Eisenhower's Executive Order 10575 of 1954 had established the Country Team, formally placing the Ambassador in charge of all civilian agencies in Vietnam. However, the different agencies guarded their independence—USAID, USIA, and the CIA each had a larger staff than the Ambassador and looked back to Washington, rather than to the Embassy, for direction and funding authorization. Every department had its own idea of how to run the war, influenced by different responsibilities and funding set by Congress. While the successive Ambassadors technically had veto power over these departments, none exercised it, so there was no unified direction for the COIN campaign. Nor was there one person or agency stateside responsible for pacification—below the President, "everybody and nobody was responsible for coping with it in the round."[4] With no single leader or organization accountable, the campaign adopted the attributes of the goliath, becoming overmilitarized.

To overcome bureaucratic resistance, President Lyndon Johnson tapped respected retired Army General Maxwell Taylor as his new Ambassador in 1964. The new President granted Taylor proconsular powers over all American personnel in the country, directing "that this overall responsibility includes the whole U.S. military effort in South Vietnam and authorizes the degree of command and control that you consider appropriate."[5] Taylor formed the Mission Council, a regular meeting of the heads of all the agencies and the military in Saigon. Although the council improved information flow, it could not impose a unified response as the Ambassador allowed each agency to appeal decisions to its Washington headquarters. Taylor also faced resistance from the new military commander, General William Westmoreland, who felt that efforts by the State Department to direct him regarding strategic matters violated "the prerogatives of the military commander."[6] The new Ambassador was unwilling to put Westmoreland "in the unhappy position of having two military masters"[7]—himself and the Commander in Chief of Pacific forces.

A year into Taylor's tenure, little had changed. There were now 60 poorly integrated programs under the U.S. mission. Senior officials in Saigon still looked to their departments in Washington for direction, and personnel in the provinces established independent offices with no unified chain of command. Complicating matters, deployed representatives received conflicting directives from Saigon and the United States.

As civilian agencies pursued their own programs, the military dramatically increased its ground presence in 1965. Under Westmoreland, American forces shifted away from efforts to build up the government of South Vietnam (GVN) and ARVN and instead assumed the primary combat role, engaging the North Vietnamese Army (NVA) and the Viet Cong (VC) in large-scale operations. The Army commenced its Search and Destroy tactics, embracing firepower and attrition rather than intelligence and population security to defeat the insurgents. Late in the year, Henry Cabot Lodge replaced Taylor, and reiterated Johnson's focus on pacification but did little to advance civil-military integration—in a January 1966 conference with senior Washington officials, Lodge and the other departmental directors from Saigon expressed confidence in command arrangements, and urged policymakers to maintain separate lines of communication back to individual departments in Washington. The conference produced no agreement on how to improve pacification and reminded political leaders that any major reorganization would wallow without presidential intervention.[8]

A month later, Johnson instructed the Saigon mission to produce goals and deadlines for its pacification plans. Meanwhile, other administration officials called for a single director of pacification. Heeding this advice, Johnson directed Deputy Chief of Mission William Porter to lead all pacification efforts, to the chagrin of Lodge. However, Porter did not embrace his new power, and saw the job primarily as a coordination effort. He noted that he had no desire "to get into the middle of individual agency activities and responsibilities," and sought "to suggest rather than criticize."[9] Washington proved no more receptive to presidential direction, as departments resisted the administration's major initiatives throughout 1966.[10] Nor did the military subscribe to the new COIN emphasis; generals in Washington and Saigon ignored the PROVN study,[11] which recommended that the Army shift its focus from conventional operations to a new pacification campaign directed by a unified chain reaching from a single head in Washington down as far as the district level.

The Arrival of Komer.

In March 1966, Johnson appointed Robert Komer as Special Assistant to the President for Affairs in Vietnam, granting authority to supervise and direct all U.S. nonmilitary pacification programs from Washington. The gruff "Blowtorch Bob," formerly a presidential aide, was intolerant of bureaucratic inertia, an attitude which would help him force departmental integration. In the meantime, the military centralized its COIN efforts under the Revolutionary Development Support program to encourage lower levels to focus on pacification. In a limited reprise of his 1964 edict to Taylor, Johnson gave Lodge authority to "exercise full responsibility" over the advisory effort—this time, the Army's main battle units would remain independent.[12] Although Westmoreland cooperated, the disparate civilian agencies and the military were unable to coordinate effectively.

Frustrated with the lack of progress, in August 1966, Komer suggested three possible structures to reinvigorate pacification efforts, with varying models of intercivilian and civil-military integration. The civilian agencies rejected the plans, offering no alternatives. Defense Secretary Robert McNamara proposed consolidating all pacification personnel and programs under Military Assistance Command-Vietnam (MACV),[13] with a deputy for pacification in charge; only Komer and the Joint Staff endorsed this, as other departments again demurred, though this time they offered counterproposals. USAID preferred a structure of multiagency committees at each level of the chain, creating coordination but not true integration. Under Secretary of State Nicholas Katzenbach offered yet another conservative alternative which maintained separate military and civilian chains of command.

McNamara summarized the frustration of reform advocates—"from the point of view of the important war (for the [support] of the people)—no better, and if anything worse off."[14] He continued that "[w]e have known this from the beginning. But the discouraging truth is that, as was the case in 1961 and 1963 and 1965, we have not found the formula, the catalyst for training and inspiring [the Vietnamese] into effective action."[15]

As a compromise short of civil-military integration, Johnson directed the establishment of the Office of Civil Operations (OCO) in November 1966. Over the objections of the agencies, this combined all civilian pacification operations under one chain of command reporting to the Deputy Chief of Mission; military COIN efforts remained separate. Significantly, OCO consolidated all civilian field personnel in multiagency compounds, which provided their MACV and South Vietnamese counterparts with a single point of contact. The OCO chain wrote performance evaluations for its personnel, though the parent agency added comments. Pay and administrative support, however, remained with an individual's parent agency, as did funding, which meant that OCO was unable to redistribute funds along functional lines.

Johnson gave OCO about 4 months to prove its effectiveness. The organization fell short, showing little progress against the VC. With no integration into the military chain of command, the new institution could not enlist the vast resources of MACV.

Komer noted that the peacetime approach of the deployed departments to funding, resources, and personnel was insufficiently flexible to meet the demands of COIN.[16] The lack of unified management (notwithstanding OCO's recent but short-lived attempt to unify civilian efforts) hampered the American response to the insurgency since an orthodox bureaucracy pursued multiple plans detached from the Vietnamese government. Poorly coordinated U.S.-GVN efforts inhibited the development of the Vietnamese government and military; in its stead, MACV took over the war, which undermined local ownership.

CORDS—"CAN OCO REALLY DO SOMETHING?"[17]

Civil-Military Integration, At Last.

Dissatisfied by OCO's inability to produce quick results, President Johnson signed National Security Action Memo 362 on May 9, 1967, which declared that all pacification efforts would fall under MACV, to be implemented by a civilian deputy with the rank of Ambassador.[18] To fill this role, Robert Komer would depart for Vietnam to direct the Civil Operations and Revolutionary Development Support (CORDS)[19] program. Komer served Westmoreland not as an adviser or a coordinator, but as a component commander with all that position's requisite authorities. As Khalilzad's close relationship with the White House increased his influence in Kabul, CORDS benefited from Johnson's well-known confidence in Komer.

The combined organization sought to integrate the civilian and military pacification efforts. Previously, the military had neglected pacification, contending that it was primarily a civilian responsibility. However, the military controlled most of the resources and forces. Johnson and Komer felt that pacification was too large a task for civilian agencies to handle alone, and that the effort was failing in large part because of poor population security, which could be improved only if responsibility was assigned to the military. Furthermore, they maintained that the tasks of COIN required departmental integration so tight that traditional government cooperation would be insufficient. By combining all programs under the military but with a civilian leader, CORDS created a "unique, hybrid civil-military structure which imposed unified single management on all the diffuse U.S. pacification support programs and provided a single channel of advice at each level to GVN counterparts."[20] Komer praised new Ambassador Ellsworth Bunker for overcoming resistance from civilians who resented subordination to the military, and for pressuring Army leaders who did not want responsibility for "the other war."[21]

In the new structure, Komer reported to Westmoreland as a 3-star equivalent. For the first time in the war, a single person commanded all soldiers and civilians involved in COIN. This arrangement was unusual—it was rare for civilians to operate within a military chain of command, and Komer was the first official of ambassadorial rank to serve under a military superior and have command of military personnel.[22] In addition to the benefits of unity of command, Komer envisioned the new organization as a way to give the military portion of pacification a civilian bent and interagency access to military resources.

Komer took several steps to reduce civilian discomfort with the apparent military take-over. First, he depicted the merger as one in which OCO absorbed the military program Revolutionary Development Support (RDS), and ensured that the new acronym began with the word "Civil." Additionally, to stem resignations, he emphasized the new opportunities for leadership now open to civilians. Bureaucratically, he ensured that his own CORDS staff within COMUSMACV was not simply advisory, but rather an operational group with command authority; the staff also gave CORDS considerable planning capacity. Finally, he expected and received the prerogatives of a 3-star officer, which in turn conferred legitimacy upon the program.[23]

CORDS took control of all civilians involved in pacification, including personnel from State, USAID, CIA, USIA, the White House, and the military services. Civilian and uniformed personnel were interspersed throughout the new chain of command, with military reporting to civilian, and vice versa. Komer based hiring for the organization's new positions on capability, considering those in and out of uniform.[24]

Vertical Organization.[25]

Under Komer, regional deputies for CORDS (DEPCORDS) served in each of the four Corps headquarters, with the same relationship to the Corps commander that Komer held with Westmoreland. The regional deputies, usually civilians, received a 2-star equivalent rank, and served not as advisors to the Corps commander, but rather as a component commander, with access to the Corps resources. Guided by tasking similar to Komer's, DEPCORDS were responsible for all pacification operations, "supervising the formulation and execution of all military and civilian plans, policies, and programs" in the region.[26] On pacification matters, Komer insisted that DEPCORDS report straight to him, which suited the Corps Commanders who preferred to concentrate on the conventional war.[27] The four regional deputies[28] directed the provincial teams within their areas.

A Provincial Senior Advisor (PSA) led each of the 44 province teams. The DEPCORDS, with the approval of Komer and Westmoreland, appointed each advisor within his region, ensuring that the high command of CORDS closely controlled personnel selection. Roughly half the PSAs were civilian with military deputies (DPSA); uniformed Provincial Senior Advisors had civilian deputies. The PSA was in charge of all pacification in the province, including American military advisors attached to local Vietnamese militia (the Regular Forces and Popular Forces [RF/PFs], Revolutionary Cadre, and the Home Guard). By detaching militia advisors from their counterparts who supported ARVN conventional forces, CORDS freed them from the influence of orthodox tactics. This move also enabled the local militia to concentrate on securing the population rather than supporting the ARVN. To implement his programs, the PSA worked closely with the Province Chief, usually a colonel in the South Vietnamese military. The Province Chief supervised the provincial government and militia, as well as the local constabulary forces, the RF/PF.[29]

Every province had several district teams, each reporting to its PSA. The teams advised the District Chief, a Vietnamese official in charge of the RF/PF and the Revolutionary Development Cadre, on pacification and development. The country was covered by 250 district teams (see Appendix 3 for map with districts and provinces).

The new structure survived some initial bureaucratic challenges. Corps commanders objected to elements of the reorganization, but Westmoreland supported the CORDS chief—although the commanding general's search and destroy methods had been part of the problem, he did not impede this solution. In Washington, USAID officials preferred to use their own Saigon channels rather than CORDS for pacification matters. The USAID Administrator appealed to Under Secretary of State Katzenbach to remove the agency's programs from CORDS, but Komer successfully countered that basing the pacification structure on administrative lines to Washington had impeded the necessary civil-military unity.[30]

Team Structure.

About 100-125 personnel lived on a provincial CORDS compound. Serving alongside 100 soldiers, 12 civilians assisted in providing security, aviation support, administration, communications, logistics, and advising. Under the PSA and DPSA, an Executive Officer managed personnel. The team was divided along functional lines—most provinces had New Life Development (NLD), Public Safety (police training), Refugees, Psychological Operations (psyops), and Chieu Hoi.[31] An NLD Chief ran all USAID programs, including community development and agricultural assistance, with the help of two or three field assistants. An American police officer or soldier would serve as Public Safety Advisor. The military usually staffed administrative positions and psyops. Medics were military or civilian.

Each district team, comprising four or five officers and NCOs (and, in calmer districts, civilians), reported to the provincial team. The district teams lived on the District Chief's headquarters compound and worked closely with the RF/PF and Revolutionary Development Cadre. The RF/PF, a local militia, provided security for villages and hamlets. The Revolutionary Development Cadre combined development and security, with tasks varying from building schools to attacking VC infrastructure (through identification, coopting, capture, and offensive operations). Including district teams, the PSA was responsible for 140-170 personnel.

ASSESSMENT BY PARTICIPANTS

Chain of Command.

Unity of command was complete and effective throughout the new pacification organization, with CORDS seniors accountable for the performance of their geographic areas and granted authority over subordinates, no matter the individuals' parent agency. Below the Corps, the PSA "ran the show,"[32] with members of the province and district teams unambiguously under his command. The DPSA served below the PSA and was responsible for many of the daily operations. All personnel reported to the DPSA, most through the Executive Officer (XO) and the New Life Development Chief. A uniformed XO was responsible for the soldiers performing support functions; the NLD Chief managed all development personnel. A civilian DPSA described the command system—"I had the exact same level of command over civilians as military."[33] He noted that when CORDS began, there was a difficult adjustment period, citing two examples. His military deputy (the XO), a West Pointer and Ph.D. 10 years his senior, was initially uncomfortable with the arrangement, but "came to understand that I was a bit smarter than he in what we were actually doing [pacification]."[34] One level lower, a district advisor (who also reported to the DPSA) contended that civilians should not exert command over the military except at the level of the Secretary of Defense, but eventually accepted the deputy advisor's new authority.

The integration of operational control led to intimate involvement by each community in the other's activities. A USAID representative who served as Deputy PSA appeared unannounced in the districts to monitor the night patrols of the advisory teams. Although all members of his five districts were military, he noted that they respected the chain of command, and approached him for support as well as directives.[35] Another DPSA, armed with an M-1 carbine but no prior military experience, joined district teams on night patrols and ambushes. As "the senior leadership, I wanted to see what was happening."[36] By most accounts, the PSA's unity of command produced tight provincial teams. A refugee advisor noted that "in no uncertain terms, the military and the civilians realized that they worked for one man, and as a result of this, I think everybody was working on the same team. Everybody got along very well."[37]

Rating authority was also shifted from the parent agency to the provincial team. The PSA wrote the DPSA's evaluation and endorsed those of the leadership team; the DPSA (a civilian if his senior was military, and vice versa) in turn wrote the evaluations of the district advisors, the XO, and the NLD chief, and endorsed the evaluations of these leaders' subordinates.

Performance in the Field.

CORDS generally earned high marks for improved efficacy relative to its antecedents. A USAID participant noted that the new program was better structured, with programs more responsive to the economic needs of the people. After continual changes to strategic plans in an earlier civilian-only program, he complained to Saigon, "Why don't you settle on a plan and we can give it a chance to work?" CORDS worked "much, much better."[38] He praised the increased access to resources, particularly at the Corps level, noting that previous programs had no levels of support between the field and Saigon. Another CORDS participant described the local flexibility of the program, noting that Komer's staff specified only the chain of command, certain functional sections, and a presence at the district level, but left subordinates free to adjust the organization to the circumstances. He also praised the quickened response time, noting that the integration of the new organization at the local level allowed redevelopment to commence immediately after the militia had cleared a village of VC.[39]

The unified chain of command fulfilled Komer's vision of a pacification campaign marked by civilian tone and fueled by military might. A USAID DPSA reflected, "We all wanted to win the war. But most civilians thought the military didn't know how to win. We were concerned that the civilian program would be submerged—that didn't happen. In many ways, the civilian side was enhanced because they now had military guys who had skills that they wanted to be used."[40] One PSA described the resource difference as an order of magnitude, noting that OCO personnel who previously had a dozen trucks at the provincial level now had access to hundreds of vehicles from a division; when organizing working parties for skill-intensive jobs, instead of dozens of men available to OCO, CORDS could now access entire battalions. He continued, "That's one of the great things that came, I think, out of CORDS. Now those who were in civil planning-type jobs could feel much more free about demanding, not just asking or begging for—but demanding military resources and expect to have them made available to them."[41]

Problems.

The most frequent complaints described onerous new reporting requirements which detracted from field work,[42] although one participant appreciated the elimination of dual reporting to OCO and MACV chains.[43] Others objected to the militarization of USAID programs[44] or to the consolidation of power on the civilian side of the program.[45] Some noted strain between the civilians and uniformed in the new chain of command.[46] Finally, several observed the tension between USAID personnel assigned to Saigon and those deployed with CORDS. A USAID officer in Saigon described the difficult coordination between long-term national programs run from the capital and local programs delegated to CORDS.[47] Those in the field concurred—a CORDS Rural Development Officer contended, "One does not advise the other as to what they are doing. In other words, how can a person in CORDS create what I call lasting institutions when they have no idea what the long-range economic planning is at the Saigon level, so rather than working parallel to one another they work perpendicular to one another."[48] He concluded that conflicting local and national programs impeded CORDS's cooperation with the local government.

EFFECTS AND EVALUATION

Results.

CORDS fundamentally changed the American and Vietnamese approach to the insurgency. The new organization made the military accountable for pacification, eliminating prior ambiguities which had let generals claim that COIN was primarily the responsibility of civilian departments. MACV's list of priorities reflected the new emphasis, as pacification rose from seventh to second of seven major goals.[49] The unified approach to COIN bestowed prestige on the American and South Vietnamese efforts as phrases such as "non-military actions" and "the other war" fell from the theater lexicon.[50]

More importantly, vast military resources revitalized local defense forces through arms, training, and intelligence. Improvements in rural administration, economic conditions, land reform, health, and infrastructure bolstered the local government. The budget for pacification nearly tripled between 1966 and 1970, from $582 million to $1.5 billion (with most of the increase from Defense funding), while conventional American units were withdrawing. This increased spending allowed a large increase in personnel. In 1966, the United States had 1,000 personnel assigned to pacification; by 1969, there were 7,600, including 1,200 civilians[51] (today, there are 50 in Afghanistan). In 1965, there were fewer than 100 USAID advisors; at the peak of CORDS, there were 100 agricultural advisors alone.[52]

CORDS capitalized on huge VC losses during the Tet offensive, its expansion of funding, and personnel who were directed by a unified chain of command, all effectively drained the insurgency. In early 1968, 59 percent of South Vietnamese lived in secure villages. By 1971, the figure increased to 96 percent, with most of the gains in the rural districts that had been the core of the insurgency.[53] The program achieved these results through the fundamental tenets of COIN, training 900,000 security and government personnel, including 300,000 civil servants.[54] These efforts created a viable local administration which provided adequate security and effective services, displacing the shadow VC regime that had subverted the state in much of the countryside. Sapped of popular support, the Viet Cong presence declined dramatically between 1967 and 1971.[55] Assessing the insurgency's decline, a VC colonel reflected that "[l]ast year we could attack United States forces. This year we find it difficult to attack even puppet forces. . . . We failed to win the support of the people and keep them from moving back to enemy controlled areas."[56]

Finale.

While CORDS suppressed the insurgency, the ARVN increased its capacity. As the VC lost power, CORDS shrank from 1970 to 1972. Having largely achieved its goal, the organization lost primacy as individual departments regained control over their traditional sphere of influences; by early 1973, CORDS was disbanded. In 1972, with the help of American air power and logistical might but no U.S. ground forces, the ARVN had repelled an NVA conventional invasion. But soon the patience of the American public expired, and Congress forbade any further military assistance to the South Vietnamese government. Komer concluded that "[t]he war ended as the American military had thought it would begin"[57]—the insurgency having failed, the North turned to a conventional strategy, exploiting the American departure. Hanoi violated the peace accords, and the ARVN, with no American assistance, yielded to the NVA's second attempt. Saigon fell to the communist conventional army in 1975.

Komer maintained that if pacification had been reformed earlier in the war, victory would have been more likely, though not certain.[58] Unity of command was only one of three improvements enabled by CORDS—the program also increased the scale of pacification and promoted a civilian tone. However, unified command facilitated the latter two. Without responsibility for pacification, MACV would not have increased its contribution; and without a unified chain, with management delegated to a nonmilitary head, civilians would have had little influence on the military's COIN efforts.

Why the delay in adopting an effective COIN strategy? Pacification received presidential attention as early as the Kennedy administration—Komer describes *"the striking contrast between the amount of policy stress on counterinsurgency, or pacification, and how little was actually done—up to 1967 at least"* (emphasis in original).[59] In explaining this lag between policy pronouncements and practice, Komer cited four impediments. First, institutional inertia causes organizations to change slowly and incrementally. Second, poor institutional memory, exacerbated by short tours, impeded institutional learning. Third, inappropriate incentives encouraged conformity and punished innovation. Fourth, organizations failed to analyze their performance because of a bureaucratic reluctance to self-criticize.[60]

LESSONS FROM VIETNAM

How does our Vietnam experience inform our efforts in Afghanistan? There are strategic similarities between the two conflicts. As in Vietnam, American forces face an insurgency that considers native popular sentiment (and, by extension, American public opinion) to be the center of gravity. As did the VC, the Taliban portray their cause as a struggle against a colonial occupier and its puppet. Afghan insurgents follow their Vietnamese predecessors in undermining a struggling government which has difficulty extending its power to the hinterlands.

There are also significant differences. In Vietnam, the enemy fought a popular revolt eventually united under communist ideology. Afghan violence, fueled by tribal tensions, has varied sources, as theocratic recidivists mix with anarchist warlords and drug runners. There is also a great disparity in scale. The VC thoroughly infiltrated much of the South, creating a shadow government in large parts of the country. The Taliban insurgency has never approached the size of the Communist effort. Finally, the war in Vietnam had multiple components, the insurgency coupled with the NVA and great power sponsors in its effort to overthrow the Saigon government. In Afghanistan, the insurgents are largely on their own—while Pakistan provides a critical haven for fighters, it does not threaten invasion.

However, these differences of scale and character increase the relevance of Vietnam. While the insurgency remains muted, the coalition should prevent its spread. And with no conventional opponent looming, American and allied forces have no "other war" to distract from COIN.

The Army's Reaction to the Vietnam War.

In Afghanistan, like Vietnam, the dominant military culture is not comfortable with COIN. The Army, and the nation more broadly, interpreted the experience in Indochina as the error of attempting COIN, when in fact the interagency effort had succeeded when properly organized and resourced. Following that war, the Army sought not to enhance its COIN tactics and COIN schools that it had developed late in the war; rather it expunged COIN from its doctrine and focused again on the plains of Europe. For the Army, the Vietnam War and its confused tactics had been a decade-long distraction from the main event. Before the fall of Saigon, the Army tasked General William DePuy, the commander of the U.S. Army Training and Doctrine Command, to develop a new doctrine. Looking not to the Mekong, but instead to the Sinai and the 1973 armor clashes between Israel and the Arab states, DePuy developed Active Defense to repel a Warsaw Pact invasion. Reinvigorated, the service debated the merits of the new way of fighting and soon replaced it with AirLand Battle Doctrine, which traded the conservative defense-in-depth for coordinated ground and air thrusts to interdict Eastern Bloc echelons before they reached the front. Although never employed in its theater of design, many saw the Gulf War as a vindication of AirLand Battle.

Today's generals, commissioned in the days of Active Defense and AirLand Battle, led companies and battalions during the Gulf War and imbued their Army with the supremacy of American firepower and technology, coupling their battlefield experience with the heritage of World War II. The accelerating pace of the revolution in military affairs during the subsequent decade, followed by the quick capture of Kabul and Baghdad, cemented the primacy of the new conventional military, with political and military leaders alike in awe of network-centric warfare. The events of 9/11 had brought a new type of enemy, but not a new doctrine, as many in the military and the DoD failed to acknowledge the familiar symptoms of insurgency. Only in 2004, after the Taliban and Saddam's fedayeen refused to accept the apparent triumph, did the military again consider COIN.

But Big Army still holds sway. For most of the American intervention in Afghanistan, reservist Civil Affairs colonels directed PRTs, but sometimes clashed with their combat arms counterparts for influence in the battlespace. Leadership of a team, while respected, was not the route to stars in the Army; active duty officers instead sought the traditional major command, the maneuver battalion, allegedly tailored to the demands of insurgency, but insulated from deployed civilians who are relegated to the PRTs. During 2006, Air Force and Navy officers took command of most American PRTs to assist our stretched Soldiers. That the Army and Civil Affairs ceded the PRT role when searching for a way to conserve manpower reveals the priority of the interagency COIN mission within the service.

The military is not alone in its resistance to the principles of COIN. Civilian departments, even USAID and State with their international mission, have not embraced COIN as a focus of their overseas efforts. The nonmilitary agencies have prevented the development of new teams through slow personnel deployment,[61] failed to couple their national programs with their provincial programs, and resisted unified civil-military authority or even robust Chief of Mission authority over the civilian side.[62]

ENDNOTES - SECTION IV

1. As defined by Chester Cooper, pacification is an effort to extend the writ of the central government and undermine the enemy through a combination of programs designed primarily to (1) promote security, and (2) strengthen ties between the government and people through civil programs. The ratio of the mix depends on the local circumstances. See Thomas W. Scoville, *Reorganizing for Pacification Support,* Washington, DC: Center of Military History, U. S. Army, 1999, p. 3.

2. Andrew Krepinevich, *The Army and Vietnam,* Baltimore, MD: The Johns Hopkins University Press, 1986, p. 30.

3. Deborah Avant, "The Institutional Sources of Military Doctrine: Hegemons in Peripheral Wars," *International Studies Quarterly,* Vol. 37, No. 4, 1993, p. 418.

4. Robert Komer, *Bureaucracy at War: U.S. Performance in the Vietnam Conflict*, Boulder, CO: Westview Press, 1986, p. 82.

5. *Ibid.*, p. 90.

6. Michael Weaver, *ACSC Quick-Look—Military Assistance Command, Vietnam: The Imperative of Political-Military Unity*, Montgomery, AL: Air University, Maxwell AFB, 2005.

7. Scoville, p. 9.

8. *Ibid.*, p. 17.

9. *Ibid.*, p. 26.

10. Komer, *Bureaucracy at War*, p. 97.

11. A group of young officers compiled the study, officially titled the Program for the Pacification and Long-Term Development of South Vietnam.

12. Dale Andrade and James H. Willbanks, "CORDS/Phoenix—Counterinsurgency Lessons from Vietnam for the Future," *Military Review*, March/April 2006, p. 12.

13. American military forces based in Vietnam served under this command.

14. Scoville, p. 39.

15. *Ibid.*

16. Komer, *Bureaucracy at War*, pp. 83-84.

17. Field jargon in the program's early stages. CORDS representative in discussion with author, July 20, 2007.

18. Lyndon Johnson, *National Security Action Memorandum 362—Responsibility for U.S. Role in Pacification, Revolutionary Development*, May 9, 1967.

19. The organization's title combined the names of its two antecedents, OCO and the military's RDS.

20. Komer, *Bureaucracy at War*, p. 118.

21. *Ibid.*, p. 119.

22. Scoville, p. 66.

23. *Ibid.*, pp. 64-67.

24. Ross Coffey, "Revisiting CORDS: The Need for Unity of Effort to Secure Victory in Iraq," *Military Review*, March-April, 2006, p. 28.

25. See Appendix 2 for national and province team organization.

26. Scoville, p. 68.

27. Richard A. Hunt, *Pacification: The American Struggle for Vietnam's Hearts and Minds*, Boulder, CO: Westview Press, 1995, p. 90.

28. As four RC Commanders cover Afghanistan, four Corps Commanders covered South Vietnam.

29. Pronounced "ruff-puffs."

30. Scoville, pp. 75-77.

31. Chieu Hoi, meaning "Open Arms," was an amnesty program for VC.

32. "Debrief of a Provincial Refugee Advisor—Binh Dinh, Vietnam: 1967-1969," Honolulu, HI: Asia Training Center, University of Hawaii/AID.

33. CORDS representative in discussion with author, July 20, 2007.

34. *Ibid.*

35. CORDS representative in discussion with author, July 15, 2007.

36. CORDS representative in discussion with author, July 20, 2007.

37. "Debrief of a Provincial Refugee Advisor—Binh Dinh, Vietnam: 1967-1969."

38. CORDS representative in discussion with author, July 15, 2007.

39. "Debrief of an AID Program Officer—Vietnam: 1966-1968," Honolulu, HI: Asia Training Center, University of Hawaii/AID, p. 6.

40. CORDS representative in discussion with author, July 20, 2007.

41."Debrief of a CORDS Representative—Bin Duong, Vietnam: 1966-1967," Honolulu, HI: Asia Training Center, University of Hawaii/AID, p. 30.

42. See "Debrief of a Provincial Refugee Advisor—Binh Dinh, Vietnam: 1967-1969," p. 17; "Debrief of a Military Information Specialist—Dinh Tuong, Vietnam: 1967-1969," p. 21; "Debrief of a Rural Development Officer—Vietnam: 1967-1969," Honolulu, HI: Asia Training Center, University of Hawaii/AID, p. 12; "Debrief of a CORDS Official—Quang Tin, Vietnam: 1967-1969," Honolulu, HI: Asia Training Center, University of Hawaii/AID, p. 6.

43. "Debrief of a CORDS Representative—Bin Duong, Vietnam: 1966-1967," p. 11.

44. "Debrief of a CORDS Agricultural Advisor—Kien Phong and Kien Tuong, Vietnam: 1968," Honolulu, HI: Asia Training Center, University of Hawaii/AID, p. 8.

45. "Debrief of a CORDS Official—Quang Tin, Vietnam: 1967-1969," p. 7.

46. "Debrief of a Rural Development Officer—Vietnam: 1967-1969," p. 13.

47. "Debrief of an AID Program Officer—Vietnam: 1966-1968," Honolulu, HI: Asia Training Center, University of Hawaii/ AID, p. 2.

48. "Debrief of a Rural Development Officer—Vietnam: 1967-1969," p. 13.

49. Scoville, p. 82.

50. *Ibid.*, p. 80.

51. Figures from Andrade and Willbanks, pp. 16-17.

52. Michael J. McNerney, "Stabilization and Reconstruction in Afghanistan: Are PRTs a Model or a Muddle?" *Parameters*, Winter 2005-06, p. 44.

53. Komer, *Bureaucracy at War*, p. 151.

54. *Ibid.*, p. 126.

55. Coffey, p. 33.

56. *Ibid.*, p. 31.

57. Richard Betts, "Blowtorch Bob in Baghdad," *The American Interest*, Vol. 1, No. 4, 2006, p. 35.

58. Komer, *Bureaucracy at War*, p. xv.

59. *Ibid.*, p. 136.

60. Robert Komer, *Bureaucracy Does Its Thing: Institutional Constraints on U.S.-GVN Performance in Vietnam,*, Santa Monica, CA: RAND, 1972, p. viii.

61. At the height of CORDS, 20 times as many civilians participated as currently staff American PRTs; even proportionally, PRTs, with fewer than 5 percent civilian, lag CORDS teams that were 15-20 percent civilian.

62. Consider the frequent insistence by AID employees that their agency is still independent of the State Department and they should remain so in the field. Multiple sources.

SECTION V. WHY THE SEGREGATED RESPONSE?

Civil-military integration in the American response to the Afghan insurgency has lagged for three reasons. First, the organizational culture of civilian departments differs from that of the military. Second, the executive branch has not matched the efforts of the Johnson administration to promote integration. Finally, the American societal concept of war still reserves the battlefield for the military, expecting little political interference or civilian participation at the operational level and below. In the face of these obstacles, the government has forsaken unified authority, and instead pretends that "unity of effort" is sufficient.

CULTURAL DIFFERENCES

Military culture relies on force, rewards action, embraces command, and is reassured by quantitative metrics. Command improves mission efficiency in any environment, but the other three attributes may reinforce a short-term outlook that is ill-suited for insurgency. In contrast, the civilian agencies stress persuasion over coercion, deliberation over quick decision, consensus over authority, individuality over teamwork, and effects rather than results. Some of these attributes are applicable to COIN, but all conflict with the warrior culture, and thus impede unity in theater.

The Military.

Conventional battle demands decisiveness. Fortune favors the side, whether defense or offense, which develops, rehearses, and most importantly, efficiently executes its operational plan. This does not imply rigid allegiance to the original strategy. The chaos of combat requires constant refinement based on the new realities of the dynamic environment; quick implementation of these adjustments may carry the day. But immediate action is not always as effective in COIN as in conventional combat. The irregular environment sometimes rewards a restrained commander who appraises the village alliances before attacking or apprehending purported insurgents.

Command is the most effective decisionmaking process in an environment that rewards speed over perfection. The military embraces faith-based authority in which a subordinate obeys an order, whether or not convinced by its logic, because he understands that his picture of the battlefield is less complete than that of the commander. The directive decisionmaking process of the battlefield permeates other aspects of military life, thus creating an organization comfortable with hierarchy. Recognizing that the effective waging of war depends on the fulfillment of each individual's duty in accordance with the commander's intent, the military creates an environment in which subordinate officers see themselves as members of a team and ensure that their units or areas of responsibility contribute to the overall mission of the organization. Doctrine gives each member, from private to general, a familiarity with the way to fight, and a confidence that the commander's directives will fall within accepted parameters.

Guided by the sequential decisionmaking process of battle, the commander needs data to evaluate and adjust the plan. The military culture promotes metrics as the input, and sets milestones as intermediate goals en route to victory. Conventional war is particularly suited to metrics—the American military can measure its gains by ground captured and tanks destroyed, carriers sunk, and divisions enveloped. But destructive metrics are generally more mensurable than constructive, and in insurgency, when ground is often nominally in the control of the occupation but infested with insurgents, the military must rely on more abstract measures of popular support. Against guerrillas, commanders unable to resist the simplicity of negative metrics may succumb to rebel body count as an indicator of success. Even those who maintain that less palpable progress, rather than tangible destruction, is a surer route to defeating an insurgency, may employ measurements which emphasize results (number of wells or schools constructed) rather than less easily mensurable effects (community nutritional benefits and educational levels). In sum, the culture of metrics prefers quantifiable markers to those more abstract.

Finally, the military tends to perceive campaigns in a short time frame. Missions in its comfort zone have clear goals, a finite duration, and a specific exit strategy. At operational and tactical levels, commanders and units feel that they have a job to complete by the end of a tour. When combined with the culture of action and metrics, this mindset may have a negative impact on a counterinsurgency campaign. A unit is eager to show progress during a 6- or 12-month tour through concrete results. It may initiate projects designed for short-term gain rather than long-term improvements, such as schools constructed with little regard for the local government's capacity to provide teachers, or jobs created with little attention to the effects on the local economy and village power structure.

State and Other Civilian Agencies.

As the military is accustomed to interstate conflict, the State Department is designed for traditional state-to-state relations. There are significant differences between the two departments. The Ambassador cannot ordinarily force his counterpart to yield, but rather must persuade a sovereign power to assent. Whereas battle favors decisive action, diplomacy requires deliberation over the prolonged cycle of negotiation, which not only characterizes relations with the host nation, but also the forming of alliances within the mission and with members of other embassies. Bureaucratic structure reinforces these different approaches to decisionmaking. With a small planning organization, an embassy has difficulty matching the decisionmaking pace of a task force command, since diplomats "can't keep up with a stable of 200 over-caffeinated majors."[1]

Whereas one state can force war upon another, states must *agree* to negotiate with each other. No matter the disparity in strength of the parties, there must be a consensus between each that the agreement is preferable to the status quo; if not, one state will not accede to the proposal. Because each side holds this veto power, even the best diplomat may have limited effect on an obdurate counterpart. Recognizing the inevitability of compromise and mindful of future negotiation, the diplomat recalls the historic admonition, "Above all, not too zealous."[2]

The military grooms its officers for command. The rising diplomat, with no troops to direct, demonstrates his individual excellence through the available avenues—cables to the mission and Washington, and service to the Chief of Mission. When he becomes ambassador, the diplomat does not see his power derived from the force-multiplying effect of a grand team, but rather from the impact of his skill as a negotiator, the clarity of his strategic vision, and the clout of the coalition developed through his relationships. With fewer leadership opportunities during a career, an FSO is less comfortable with command and its hierarchy, progressing instead in a system that puts a premium on relationships with foreign counterparts and embassy colleagues. While mammoth scale requires doctrine to impart common operating concepts on the military, diplomats resist a formulaic course of action that strips the art from the consensus-building that comprises their duties. To some degree, the training in negotiation prepares FSOs for the intimate and frustrating relationship with the local sovereign that characterizes COIN. But COIN is still war and, as such, requires the decisiveness enabled by command and broad management experience gained in the military.

Additionally, the success of the diplomat is less prone to measurement. Negotiations may succeed or fail, but the process and the results are only grossly quantifiable. It is difficult to attribute values to particular stages of talks—what influences actually determined the final arrangement?

Finally, the military and diplomatic cultures have different conceptions of time. While the military hopes to quickly enter and exit a conflict, diplomats seek to maintain tolerable relations with a state, a task without end. Military culture encourages short-term accomplishment, but State leaders may be more concerned about the sustainability of programs. For COIN, enduring solutions are important, as they extend the control of the local government. However, the interest of diplomats in returning to normality may undermine a COIN effort, as institutional norms promote a peacetime manning and organizational structure that is ill-suited for the demands of insurgency. In Afghanistan, several government officials noted that the recent decision to normalize the Embassy in Kabul decreased the level of integration with the military and between civilian agencies.[3]

Pink on Pink.[4]

Cultural differences also impede integration of the State Department and USAID, despite their shared heritage. In peacetime, the two agencies work closely within American missions in developing countries and both prioritize relationships with local officials. USAID has been brought under State direction over the past decade, but this subordination has occurred only at the highest levels of the organizations—the administrator of USAID now reports to the Secretary of State, and serves concurrently as the Director of Foreign Assistance (responsible for most State funds as well as USAID grants). Below the administrator, the agency functions as a separate organization from the State Department, with the USAID headquarters directing overseas projects and funding. This separation reinforces cultural differences. While diplomats may consider themselves strategic thinkers helping to set broad policies from an embassy, USAID representatives perceive themselves as implementers who feel at home when in the field.

On the provincial teams, each agency does its thing—FSOs seek to influence local political leaders, while development workers use their funding to improve infrastructure. Portions of USAID culture are also considerably influenced by the NGO ideals of humanitarian assistance, which dictate that donors deliver aid on the basis of need and refrain from using assistance as political leverage. USAID relies heavily on contractors, many of whom previously worked with NGOs, whose neutral spirit makes them less comfortable than their military or State counterparts with the coupling of local programs to a patently political mission—extending the reach and authority of the Kabul government. One employee who had spent extensive time in Afghanistan examining USAID's role on PRTs noted, "There is still a philosophical battle between old USAID and new USAID. Those who joined 20-30 years ago . . . have a hard time adjusting to how we're supporting stability operations, and how we're related to counterinsurgency."[5]

"IT DIDN'T TAKE LINCOLN THIS LONG"[6]

COIN scholar Andrew Krepinevich attributes the lack of unified authority in Afghanistan to executive indifference. His evaluation echoes the lessons of Vietnam, which suggest that true integration requires prolonged presidential attention.

The Kennedy administration recognized the nature of the conflict in Vietnam and sought to focus the government on the threat through directives to the military, an interagency task force (Special Group—Counter Insurgency), and a policy document which described the danger of insurgency and the appropriate American assistance to the host government. The "United States Overseas Internal Defense Policy" called for an integrated civilian and military effort, directed by the Chief of Mission.[7] Although they espoused the new principles of unified response, the bureaucracies successfully avoided integration. Aware that organizations advertise change while shunning true reform, President Johnson tried several times to force unity of command on all American efforts in Vietnam. He granted Taylor control over all American assets, military and civilian, in 1964, but the retired general failed to employ his power. Two years later, Johnson found another reluctant partner, Deputy Chief of Mission William Porter, who would not embrace command over the more limited COIN mission. However, the President's attention did not wane, and in Komer he found a lieutenant eager to exercise this broad mandate.

Four decades later, the government has again focused much attention on COIN. Most significantly, the State Department is promoting a Civilian Response Corps which would provide hundreds, and eventually 2,000, deployable civilian experts to assist COIN efforts. If passed by Congress, this will dramatically increase the civilian agency's capacity in COIN. Other COIN initiatives have focused on training and education. In the last year, the State Department's Bureau of Political-Military Affairs and DoD's Office of Stability Operations have jointly organized conferences, created a website, and led the drafting of "21st Century Counterinsurgency: A Guide for U.S. Policymakers," to introduce the broader government community to the challenges of COIN. DoD is leading the development of a Center for Complex Operations which will coordinate the courses each department has developed to train its personnel in COIN. The Center for Army Lessons Learned has drafted a "PRT Playbook" to educate the multiagency teams. The government has recently developed combined courses to train civilian and military personnel deploying to PRTs. These are important changes and bolster interdepartmental cultural familiarity.

The government also recognizes the importance of an integrated civil-military response to insurgency. Published jointly by the Army and Marine Corps last year, *Field Manual 3-24, Counterinsurgency,* contends that "[a]lthough military efforts are necessary and important, they are only effective if integrated into a comprehensive strategy employing all instruments of national power."[8] In Directive 3000.05, issued November 28, 2005, the Pentagon declares that "integrated civilian and military efforts are key to successful stability operations. Whether conducting or supporting stability operations, the Department of Defense shall be prepared to work closely with relevant U.S. Departments and Agencies, foreign governments and security forces, global and regional international organizations."[9] One week later, the President signed NSPD-44 directing the Secretary of State to "coordinate and lead integrated United States Government efforts" in order "(i) to coordinate and strengthen efforts of the United States Government to prepare, plan for, and conduct reconstruction and stabilization assistance . . . in a range of situations that require the response capabilities of United States Government entities, and (ii) to harmonize such efforts with U.S. military plans and operations."[10]

But these efforts are not sufficient. While the Bush administration gave considerable attention to COIN, it never reached the necessary level of commitment to a unified authority; it took several years of such executive insistence to wear down the bureaucracies and achieve unity of command in Vietnam. Instead of major in-theater organizational reforms, the government has relied on the ad hoc evolution of PRTs, allowing the departments to maintain insulated reporting chains. In Washington, bureaucratic adjustments are similarly limited—NSPD-44 conferred influence on a new directorate within the State Department, the Office of Reconstruction and Stabilization. However, this "lead" does not translate to authority or money, so State has no power to direct a unified effort to assist a struggling government. Furthermore, the Coordinator is an Assistant Secretary, limiting his influence even within the State Department where other bureaus resist the new organization. This internecine conflict diminishes the clout of the office in talks with other departments. Officials from State and Defense cite the contentious negotiations regarding security responsibilities for PRTs in Iraq, which delayed full deployment of the teams for over a year, as a product of ambiguous departmental responsibilities. Thus, despite the policy focus in Washington, neither the Presidential Directive nor departmental injunction has called for a single unified command at the theater and tactical level; in the field, segregated command still prevents integrated response.

WORLD WAR II, ITS GENERAL, AND THE AMERICAN PSYCHE

Cultural tensions between the agencies and the government's unwillingness to unify authority impede the American response to insurgency. Less apparent, but no less determining, is American society's discomfort with COIN, demonstrated by the fact that our COIN success in Vietnam has not been more widely recognized. The aversion to irregular warfare stems from three legacies of industrial-age victories, particularly World War II. First, Americans prefer the putative separation between political and military functions in conventional warfare over the intimacy between military actions and political effect called for by COIN. As part of this ideal, people imagine a form of warfare in which noncombatants are little involved. Second, the popular recollection of World War II minimizes the civilian contribution to the victory, instead attributing success to the military's freedom from political interference. Third, Americans share General MacArthur's impatience with the strictures of limited war that have replaced the World War II commitment to victory. These factors fuel the American citizen's anxiety with COIN, a quintessentially political-military form of war which blurs the lines between combatants and civilians, demands political guidance at every level, and depends not on preponderance of force but on discrimination in applying it.

The American Way of War.

Government documents and professional journals describe insurgency as primarily political, as if conventional war were not.[11] But at a macro level, conventional war is as political in its essence as insurgency. It changes the fabric of power—territorial sovereignty, control of natural resources, the fate of a regime—as much as an insurgency does. The difference between an insurrection and conventional conflict is found at the micro level. In the former, the politics is incremental and ubiquitous. The private must consider how his bearing at a checkpoint will sway the scores that pass through during his 6-hour watch, their loyalty being the prime determinant of the war. Which political competitor will they choose, the nascent state and its foreign protector or the insurgency? In orthodox war, the politics is deferred in the American mind—it falls at the end, after the fighting, at the negotiating table. The general in command need not worry about the instantaneous effects of his actions on the local population. The enemy's forces are uniformed, as are his; one army will remain on the field at the end of the day, and it is the commander's job to kill as many of the enemy army while losing as few of his own as possible, with little worry about distant civilians.

The peculiarities of America's World War II experience exacerbate this misperception of the separation between the political and military. For the United States, the fight began and ended in the Pacific, a theater named for a broad ocean, not a teeming continent. At sea, warfare mirrors the apolitical ideal. Admirals have the terrifying but unambiguous task of destroying the enemy force. The political effect—a blockade, a stranded colony—follows the naval battle, completely segregated from the throes of combat. And, with their misery distant from our shores, it is harder to remember the civilian targets, the Japanese starving as their merchant fleet sank. Recalling the ground combat of both theaters, history focuses on clashes with the enemy, with little attention to the noncombatants, often unable to escape the front, swept up in the carnage.

With news media intent on bolstering morale on the home front, Americans were spared most of the details of the civilian horror of World War II. The public was aware in the abstract of the human price borne by the enemy populations (particularly through strategic bombing), but this seemed a necessary cost in conquering the aggressor nations and ending a just war. In Vietnam, with a less pliant press, the public saw the intimate misery of the war, and it was difficult to conceive how the suffering of innocents contributed to our support of their state. The apparently antiseptic fight of the Gulf War returned the United States to the simpler political goal of expelling an invader, but with the promise of discriminating weaponry that largely spared the civilian population.

Team America.

It is not only the intimately political nature of insurgency and the confused role of noncombatants that trouble our nation. It is also the idea that war must be constrained by close political guidance and fought by more than the armed services. In the American psyche, our nation's greatest triumph was delivered by the military, free from meddling politicians. School children read of titans Patton, Eisenhower, MacArthur, Nimitz, and Arnold, leading an Army slashing towards Berlin and an invasion force hopping across the Pacific islands, both protected by a vast air armada. This victory has forged a distorted ideal of warfare—military action integrated with political guidance only at the highest level of strategic command, but segregated at the operational and tactical level. The reality, of course, was different, with Roosevelt very much involved in operations as well as strategy.

Nor was civilian participation in the war limited to the Commander in Chief and his advisers. Civilians directed the wartime economy, evaluated strategic bombing and maritime shipping, manned merchant convoys, and staffed the postwar reconstruction. But Americans have forgotten the breadth of the effort. We remember it as a triumph of arms, won by soldiers alone in the field, which, while fueled by stateside labor and industry, was neither complicated by deployed civilians nor hampered by intrusive politicians. G.I.

Joe and Rosie the Riveter are still revered, but who has ever heard of Harry Hopkins?

This conception of the military fighting abroad, supported by civilians at home, continues today. "Service" and "veteran" apply only to uniformed personnel. News programs devote clips to the sacrifice of soldiers and yellow ribbons urge Americans to "support our troops." Society still envisions campaigns waged by warriors alone; the little coverage of civilians in theater focuses on the role of contractors, the *bête noire de jour*.

MacArthur's Legacy.

According to our national narrative, the political mandate for unconditional surrender during World War II loosed the nation's might and freed our generals and admirals from political meddling. There were no significant negotiations with the Axis to constrain the Allied drives to Berlin and Tokyo. Theater commanders enjoyed latitude in their delegated authority over forces (though not to the degree popularly remembered as intra-Alliance negotiations certainly constrained the generals). To the extent that political guidance was more distant than in the past, World War II was unusual in its terms of unconditional surrender. But the Cold War, with nuclear stakes, resumed the more restrained calculus. No longer was the United States pursuing total victory; the Soviet specter dictated that political leaders intensify their control over the scope of a regional conflict lest it become global. This return to tight civilian control and constrained national means created civil-military tension since the military had been accustomed to the looser reins and vast resources of World War II. During the Korean War, MacArthur found himself unable to accept the new paradigm of limited war. After he publicly criticized Truman's decision not to bomb and blockade mainland China, the President relieved him from duty on April 11, 1951.

Upon return from Korea, he toured the nation a hero, enjoying a grand parade and rapt audiences. One week following his dismissal, MacArthur spoke before a joint session of Congress. Explaining his calls for more aggressive action against China, the general lamented the stalemate in Korea and claimed that: "Once war is forced upon us, there is no other alternative than to apply every available means to bring it to a swift end. War's very object is victory—not prolonged indecision. In war, indeed, there can be no substitute for victory."[12]

Inspiring words, certainly, but a terrifying message. For MacArthur, victory was the enlistment of national power to conquer the peninsula for the Free World, even at the risk of wider war with China and the Soviet Union. For Truman, victory was more modest, the defense of the young southern Republic and the preservation of the UN coalition. The President, after indulging the hero's hubris in the fall of 1950 and suffering China's mass entry into the war had realized that strategic myopia benighted the general's operational genius, disqualifying him from continuing in command.

The relief brought wide public and congressional criticism. Eventually, hearings on Capitol Hill judged the President's actions proper, and the nation averted a constitutional crisis. The general's luster was dimmed, but even today his advocacy of broad latitude for the theater commander, with military might unconstrained, resonates with the popular stereotype of the "Good" World War II. The Pentagon pays homage with a "MacArthur Corridor." Its center alcove, where promotion and award ceremonies are held, displays a placard with an excerpt from the "no substitute for victory" speech in 2-inch letters.[13]

Today, Korea is remembered as a draw, the opening act of America's confused entry into the era of limited war, which crescendoed with the debacle of Vietnam. A prominent narrative still blames that defeat on resource constraints and operational limits that hamstrung American forces.[14] However, the total war model—invading neighbors and pressing north towards Hanoi—may not have brought quick victory but instead simply broadened the insurgency and risked deeper Chinese and Soviet involvement. The United States lost in South Vietnam because of a misapplication, not a lack, of resources.

ENDNOTES - SECTION V

1. PRT FSO discussion with author, July 24, 2007.

2. *Ibid.*

3. David Barno and several others in discussions with author.

4. Field jargon for disagreements between AID and State.

5. AID official in discussion with author, August 2, 2007.

6. Andrew Krepinevich in discussion with author, July 12, 2007.

7. *United States Overseas Internal Defense Policy.* September 1962, p. 22.

8. *Field Manual (FM) 3-24, Counterinsurgency,* Washington, DC: Department of the Army, June 2006, p. 2-1.

9. Department of Defense Directive 3000.05, *Military Support for Stability, Security, Transition, and Reconstruction (SSTR) Operations,* Washington, DC: Department of Defense, November 28, 2005.

10. George Bush, *National Security Presidential Directive/NSPD-44,* Washington, DC: The White House, December 7, 2005.

11. And yet, news media coverage of COIN speaks of military victory, military success, and military objectives as if these are distinct from the political, and the generals might wage war insulated from the concerns of their masters and native partners.

12. Vorin E. Whan, *A Soldier Speaks – The Public Papers and Speeches of General of the Army Douglas MacArthur,* New York: Frederick A. Praeger, 1965, p. 251.

13. Author's personal observation.

14. For a discussion, see Jeffrey Record, "Vietnam in Retrospect: Could We have Won?" *Parameters,* Winter 1996-97, pp. 51-65.

SECTION VI. THE ROLE OF UNITY OF COMMAND—AN EXAMINATION OF DOCTRINE AND LANGUAGE

In the face of different organizational cultures, a lack of attention by the administration, and dated societal conceptions of war, the government seems to have made the unconscious assumption that it will not be able to bring unity of command to its various departments and has instead settled for a proxy, unity of effort. In doing so, the government has failed to apply the fundamental principle of war, unified command, to insurgency. The challenge of uniting bureaucracies is not surprising—it has long plagued the American military, and was finally addressed only late in the game by the combatant command system, which evolved over 4 decades as the government gradually shifted power from the services to a unitary military commander exercising authority over joint forces. But, because the United States is still reluctant to accept insurgency, with its intimate political nature and combined civil-military response, as war, the government has not applied the lessons of interservice management to multiagency war. While contemporary military doctrine stresses the need for unity of command to achieve unified effort, guidelines for multiagency work substitute the language of friendly persuasion for directive, embracing an elusive unity of effort without the requisite unity of command. Rather than apply the military's joint model across U.S. Government departments, policymakers have surprisingly settled for a coalition, or mutual-consent model, with its accompanying inefficiencies, even though all the members work for the same leader.

UNITY OF EFFORT—THE FRUIT OF UNITY OF COMMAND

The Path to Unified Authority within the Military.

American governments have frequently attempted to impose common direction on the disparate elements of its military. Colonials struggled to unify the action of the Continental Army and militias, and the ideology of devolved power plagued the Confederacy as governors refused to relinquish control of their forces to Richmond. Interservice disputes hampered American efforts in World War II. To address these challenges, the 1947 National Security Act began a gradual shift in power from the services to the Chairman of the Joint Chiefs, and eventually the Combatant Commanders (CCDRs), culminating with the Goldwater-Nichols legislation of 1986.

In the early 1980s, a series of poor military performances inspired the Goldwater-Nichols Act. Inadequate interservice preparation hamstrung the Desert One hostage rescue mission; confusion about command responsibility for the Marine Barracks in Beirut surfaced in the investigation following Hezbollah's attack; and a lack of common communications and doctrine impeded operations in Grenada. Among other measures, Goldwater-Nichols increased the power of the CCDRs relative to the services by granting more control of budget and training, and clearly asserting the commanders' responsibilities and authorities.

The act is clear about the importance of unity of command and its inextricably linked authority and responsibility. It asserts the "intent of Congress" as being:

> to place clear responsibility on commanders of the unified and specified combatant commands for the accomplishment of missions assigned to those commands, to ensure that the authority of the commanders of the unified and specified combatant commands is fully commensurate with the responsibility of those commanders for the accomplishment of missions assigned to their commands.[1]

To accomplish missions delegated by the President and the Secretary of Defense, the act gives each CCDR authority to direct his subordinate commands, no matter their service composition, in all aspects of joint operations, joint training, and logistics. Furthermore, it establishes that the CCDR has the authority to alter the command structure as he sees fit and suspend subordinates.

The act vests "planning, advice, and policy formulation" authority in the Chairman of the Joint Chiefs.[2] Through this power, the Chairman has developed Joint Publications, a series of doctrinal statements which serve as authoritative guidance for all members of the military regarding how to fight; "this doctrine will be followed except when, in the judgment of the commander, exceptional circumstances dictate otherwise."[3] This joint doctrine supersedes that of the individual services.

Joint Doctrine and the Principle of Command.

Joint doctrine codifies operational practices and warfare philosophy with command directives. *Joint Publication-1 (JP 1), Doctrine for the Armed Forces of the United States,* is the "capstone publication for all joint doctrine, presenting fundamental principles and overarching guidance for the employment of the Armed Forces of the United States."[4] It leaves no ambiguity about the importance of command in military performance, as well as its consolidation under a single authority. JP 1 asserts that "inherent in command is the authority that a military commander lawfully exercises over subordinates including authority to assign missions and accountability for their successful completion."[5] While a commander may delegate authority, he retains responsibility for fulfilling missions.

To integrate the varied capabilities of the Armed Forces, JP 1 describes the unity of command exercised by the CCDRs or assigned Joint Task Force Commanders (JFCs), which means that "all forces operate under a single commander with the requisite authority to direct all forces employed in pursuit of a common purpose."[6] The first principle of a Joint Force organization is simplicity—"[u]nity of command must be maintained through an unambiguous chain of command, well-defined command relationships, and clear delineation of responsibilities and authorities."[7] In practice, the military achieves unity of command by "establishing a joint force, assigning a mission, or objective(s) to the designated JFC, establishing command relationships, assigning and/or attaching appropriate forces to the joint force, and empowering the JFC with sufficient authority over the forces to accomplish the assigned mission."[8]

Unity of Command Begets Unity of Effort.

Doctrine asserts that for joint military operations "[t]he purpose of unity of command is to ensure unity of effort under one responsible commander for every objective."[9] Unified authority gives a commander the power to direct personnel and resources to a common goal. Since unity of effort is the direct consequence of unity of command for interservice operations, the doctrine stresses the primacy of unity of command, accepting it as one of nine "Principles of War" on the third page of JP 1; there is no need to list unity of effort, since it is subsumed in the master principle, and it does not appear.[10]

As unity of command is the surest route to unity of effort, joint doctrine notes its desirability outside military operations but concedes the task is daunting. JP 1 warns that during "multinational operations and interagency coordination, unity of command may not be possible, but the requirement for unity of effort becomes paramount."[11] Whereas unity of command is the *sine qua non* of unity of effort in joint military operations, the publication leaves readers little indication how to achieve that unity of effort without the "requisite authority to direct all forces" that joint military commanders enjoy.[12]

UNITY OF EFFORT—THE BEST HOPE IN ABSENCE OF UNITY OF COMMAND

The Multiagency Alternative.

To address the heterogeneous nature of war, the Joint Staff has published *Joint Publication 3-08, Interagency, Intergovernmental Organization, and Non-governmental Organization Coordination During Joint Operations*. The document reveals how the military, as the dominant player in counterinsurgency, views its relations with the other departments, as well as international actors. While JP 1 stresses that "unity of command is central to unity of effort,"[13] JP 3-08 ignores unity of command; instead the interagency publication relies on coordination, cooperation, and harmonization to achieve unity of effort.

JP 3-08 accepts the fundamental premise of sound strategy—the first page declares that "the integration of U.S. political and military objectives and the subsequent translation of these objectives into action have always been essential to success at all levels of operation,"[14] and describes the components of this integration as the "diplomatic, informational, military, and economic instruments of national power."[15] How does the military expect the U.S. Government to integrate these tools? The answer can be found in JP 1, through unity of effort. But absent are the "requisite authority" and the "unambiguous chain of command" that JP 1 cited as the key components for unity of command in achieving "common purpose." In the capstone publication, this common purpose is synonymous with unity of effort.[16] Instead, in JP 3-08, unity of effort becomes a means as well as an end, and the publication claims that "unity of effort in an operation ensures all means are directed to a common purpose."[17] While JP 3-08 does not define unity of effort, the DoD's *Dictionary of Military and Associated Terms,* part of the same doctrinal series published by the Joint Staff, unity of effort is the "coordination and cooperation toward common objectives, even if the participants are not necessarily part of the same command or organization."[18] Combine the injunctions in the preceding two sentences, and the circular logic becomes apparent: "coordination and cooperation toward common objectives . . . ensures all means are directed to a common purpose."[19] How can mere "coordination and cooperation" act to "ensure" any outcome if members refuse to "coordinate and cooperate?"[20]

Unity of effort is the route to integration in both joint and multiagency operations; but, in the absence of unity of command in the latter, what is the route to unity of effort? It seems to be mutual understanding and good will. According to JP 3-08, "coordinating and integrating efforts between the joint forces and other government agencies, IGOs, and NGOs, should not be equated to command and control of a military operation."[21] JP 3-08 recognizes that the cultural gaps between departments, including different goals, procedures, and decisionmaking processes, impede unity of effort.[22] To bridge these, this manual maintains that "close, continuous interagency and interdepartmental coordination and cooperation are necessary to overcome confusion over objectives, inadequate structure or procedures, and bureaucratic and personal limitations. Action will follow understanding."[23] The reader wonders how far action lags understanding and whether coordination and cooperation will be sufficient to conquer obstacles which, within the military, required the unified authority of the Goldwater-Nichols interservice reforms.

Differences in Joint Publication Terminology.[24]

Contrast the directive terms in JP 1 with the hortative language of JP 3-08. JP 1 cites "unity of command" 23 times, usually in the context of unity of effort. The same phrase appears twice, and only tangentially, in JP 3-08. "Authority" appears 350 times in JP 1; in JP 3-08, only 73. "Responsibility" and "responsible" appear nearly 250 times in JP 1; in JP 3-08, fewer than 100. "Accountability" or "accountable" appears 9 times in JP 1; JP 3-08 never uses either. In the place of these words, JP 3-08 uses softer terms. "Coordination" and its derivatives appear nearly 400 times, or four times per page and at twice the rate as JP 1. JP 3-08 cites "consensus" 12 times; the word never appears in JP 1.

Foreigners.

The inclusion of intergovernmental organizations (IGOs) and NGOs in large part explains the absence of directive language in JP 3-08. The American government has different constituencies, goals, and mandates than do Save the Children, the Red Crescent, or UNICEF, and no authority over most IGOs and NGOs. But why are other U.S. departments lumped with these international actors? All U.S. departments work for the same executive, represent the same nation, and strive for the same success in theater—one would expect that interdepartmental integration would be easier than its international counterpart. But so far as direction by the military is concerned, the rest of the United States government is as foreign as IGOs and NGOs.

The complicated title, "Interagency, Intergovernmental Organization, and Nongovernmental Organization Coordination During Joint Operations" is not simply a reflection of the military's attitude towards other American departments. More broadly, it suggests the difficulties of coordinating operations with other departments when there is no formal system of unified authority analogous to the Combatant Command, to direct civil-military operations. The Director of the Joint Staff, a 3-star general, signed this document. It is widely disseminated through the government, since it informs civilian agencies what to expect in operations with the military, and has suffered no backlash. The lack of displeasure indicates that while government and outside observers, and the very doctrine itself, recognize the need for integration between all departmental efforts, officials in the military and the civilian agencies have conceded that unified action between various branches of the American government presents the same difficulties as integration with international actors. Moreover, by ignoring the lessons of CORDS presented 4 decades ago, they have forsaken unified authority, the surest path to unity of effort.

JP 3-08 Terms and the Broader Government.

Other national security guidance governing counterinsurgency parrots the soft language of JP 3-08. It calls for more cooperation, closer coordination, and greater harmonization. On multiagency management charts, dotted lines of cooperation replace the solid lines of authority that characterize the diagrams of individual departments (civilian and military alike); ovals replace rectangles. The government assigns a "lead agency" but fails to define the accompanying powers, which in reality seem few. This label grants neither funding nor control—one official described it as conferring "all the responsibility and none of the authority."[25]

NSPD-44 (for the whole government) and DoD Directive 3000.05 (for DoD, uniformed and civilian, in its interaction with other departments) are the primary directives which guide agencies as they struggle to adapt to the forgotten form of COIN warfare. The President signed NSPD-44 in December 2005 to "promote the security of the United States through improved coordination, planning, and implementation for reconstruction and stabilization assistance for foreign states and regions at risk of, in, or in transition from, conflict or civil strife."[26] This mission is similar to, though broader in scope than, the PRTs' mission. The Directive assigns the Secretary of State two tasks: to "coordinate and lead integrated United States Government efforts, involving all U.S. Departments and Agencies with relevant capabilities, to prepare, plan for, and conduct stabilization and reconstruction activities"; and to "coordinate such efforts with the Secretary of Defense to ensure harmonization with any planned or ongoing U.S. military operations across the spectrum of conflict."[27] But to achieve this two-pronged mission, the directive gives the Secretary little real power, instead assigning a list of tasks to "coordinate" with other agencies. "Coordinate" and cognates appear 24 times; "authority" appears thrice (only in a final paragraph of caveats), "responsibility" and related words four times, and "accountability" once (with no uses of "accountable"). The document does describe a "lead" and "supporting" model for different agencies in particular missions, but provides no detail on what powers and duties these labels confer.

If the President himself is not comfortable subordinating departments to real authority, we should not expect the broader government to do so. DoD 3000.05, published a few months before JP 3-08, prefigured the Joint Publication's consignment of the multiagency community to the foreign realm. The 11-page directive uses the phrase "other U.S. Departments and Agencies, foreign governments and security forces, International Organizations, NGOs, and members of the private Sector,"[28] 11 times, noting that Defense personnel should cooperate, form civil-military teams, draft plans, share information, and train with this conglomerate. Only in intelligence-sharing, schooling, and personnel support does the memo distinguish the relations between executive departments from relations with international entities.

The point of this analysis is not to question laudable efforts to coordinate overseas operations with extra-governmental actors; rather, it is to demonstrate that the uniformed military is not alone in its low estimation of the prospects of multiagency integration, as all ignore the fact that every department works for the same country, the same government, and the same President. Cooperation and coordination are appropriate routes towards coupling American power with that of its allies, IGOs, and NGOs, which may have complementary but distinct goals. However, that the American government should view these same elements of loose collaboration as the main means to unify multiagency action demonstrates that officials have lost hope for the tight command structure which was so successful in CORDS. That the government has finally accepted an appropriate system, unified authority, to address similar problems in the military makes the current intransigence all the more surprising.

Of course, the concept of authority is not unique to the military. The only instance where the verb "direct" is used imperatively in NSPD-44 relates to the Secretary of State's power over the Coordinator for Reconstruction and Stabilization—all accept that the Secretary's power includes this authority over subordinates within his department. Although the Ambassador's Chief of Mission authority over other agencies is ambiguous, all Foreign Service Officers know that the Ambassador is their boss. USAID field officers have the same perspective toward their country director.

While no organization is as imbued as the military with the concept of command and its mortal implications, all institutions have its basic qualities. Perhaps the most important function of an organization is to provide a decisionmaking mechanism for its constituents. From decision derives action. The military, teams, governments, and businesses have all grasped the value of unified authority. Athletes follow coaches, or risk sitting on the bench; employees obey the directives of their bosses, lest they miss a promotion or lose their job; representatives contravene voters at electoral peril. Although not as strict as the military, and not subject to the UCMJ, all of these organizations accept the fundamental tenet of management—authority to direct action (whether creating a new business division, crafting a game plan, or passing legislation) coupled with responsibility for its effects.

ENDNOTES - SECTION VI

1. Goldwater-Nichols Act, 1986.

2. *Ibid.*

3. Letter of Introduction, *Joint Publication (JP) 1, Doctrine for the Armed Forces of the United States,* 2007, available from *www.dtic.mil/doctrine/jel/new_pubs/jp1_0.pdf,* p. A-1

4. *Ibid.*

5. *Ibid.,* p. xv.

6. *Ibid.,* p. IV-1.

7. *Ibid.,* IV-19.

8. *Ibid.,* II-3.

9. *Joint Publication (JP) 3-0: Joint Operations,* 2006, available from *www.dtic.mil/doctrine/jel/new_pubs/jp1_0.pdf,* p. A-2.

10. JP 3-0, p. I-3.

11. *Ibid.*, p. VI-1

12. JP 3-0, p. A-2.

13. JP 1, p. xvi.

14. *Joint Publication 3-08: Interagency, Intergovernmental Organization, and Nongovernmental Organization Coordination During Joint Operations,* Vol. I, 2006, available from *www.ndu.edu/ITEA/storage/728/Joint%20Pub%203-08%20Volume%201%202006.pdf,* p. I-1.

15. JP 3-08, p. I-1.

16. JP 1, p. IV-19.

17. JP 3-08, p. I-12.

18. *Joint Publication (JP) 1-02: Department of Defense Dictionary of Military and Associated Terms,* 2007, available from *www.dtic.mil/doctrine/jel/new_pubs/jp1_02.pdf,* p. 564. Note that the definition contradicts the JP 1 premise of unity of effort, in which it is a result of unity of command; in the dictionary, it can be achieved without unified authority.

19. JP 1-02, p. 564; and JP 3-08, p. I-12.

20. JP 1-02, p. 564; JP 3-08, p. I-12.

21. JP 3-08, p. viii.

22. JP 3-08, pp. viii-ix.

23. *Ibid.*, p. I-5.

24. The two publications are close in length; JP 1 is 155 pages, while JP 3-08 is 103 pages.

25. DoD official in discussion with author, July 9, 2007.

26. George Bush, *National Security Presidential Directive/NSPD-44*, Washington, DC: The White House, December 7, 2005.

27. *Ibid*.

28. Department of Defense Directive 3000.05, *Military Support for Stability, Security, Transition, and Reconstruction (SSTR) Operations*, Washington, DC: Department of Defense, November 28, 2005.

SECTION VII. PROSPECTS

Although most senior actors have, in the spirit of JP 3-08, dismissed the possibility, unifying civil and military authority in a single chain would address many of the deficiencies of PRTs and the broader American effort in Afghanistan. As indicated by JP 1, the lack of unity of command in Afghanistan has diminished unity of effort. The absence of CORDS-like integration is especially surprising in areas where there is no conventional fight to distract from COIN as it did in Vietnam. PRT and country team members recognize that the chain of command is not integrated, but, having forsaken the possibility of unified authority, cite advantages of the stovepiped status quo.

Collocation is the greatest benefit brought by PRTs, and policymakers should not minimize its significance—common quarters, shared workdays, and a strenuous environment forge strong bonds among members that bridge many of the cultural differences and foster cooperation between departments. But collocation alone does not remedy the problems of remote management or diminish the role of personalities; nor does it enlist the concepts of accountability and authority to make our war effort more effective. Moreover, our leadership in Kabul no longer enjoys the benefits of collocation.

Insurgency Alone.

In Vietnam, the large conventional aspect of the war delayed the adoption of a COIN strategy and unified chain of command. Scholars Andrew Krepinevich and John Nagl cite the Army's orthodox bias (favoring heavy firepower, concentrated units, and body count over light infantry, dispersed units, and population security) as an impediment to a proper COIN approach.[1] However, considering the strength of the North Korean and Chinese conventional forces in the previous war, it would have been careless for South Vietnam and the United States to neglect to prepare for an invasion by the NVA. This recognition of the role for conventional forces in the defense of South Vietnam does not imply that the U.S. Government achieved the proper management structure; it did not. Ideally, the President would have united all efforts under a single civil-military chain of command, as Johnson attempted during Taylor's tenure. Ultimately CORDS was a partial solution, properly integrating all COIN efforts but leaving the vast conventional mission segregated from COIN.

While it was appropriate to prepare a portion of the ARVN for conventional war, there is no analogue in Afghanistan today. COIN is the only fight. The Taliban and its associates are sufficiently weak that the Afghan military will not have to face large-scale battle; nor do neighbors threaten invasion as did North Vietnam. The unitary nature of the war makes a unified civil-military chain of command all the more imperative. Moreover, as COIN is intimately political and inextricable from the native population, the effort should probably be guided by a civilian. However, the United States is no closer to unified authority than it was in the early 1960s, approximately 5 years before CORDS was finally implemented. Field management has not received the attention from Washington that it did during the Vietnam War. Not surprisingly, practitioners and doctrine writers seem to assume that policymakers will fail to integrate civilian and military authority.

The Concerns about Unified Authority.

There is a tendency among participants to concede that a literally integrated chain would be ideal, but to defend whatever level of integration they have experienced as the highest level feasible. General Barno advocated clear authority and accountability within the military chain of command through the geographic assignments of the Regional Commands but, describing his arrangement with Ambassador Khalilzad, maintained that at the national level the military and civilian chains should remain separate—"the way we had it is as close to integration as we would want to get. . . . We're never going to find a model where the military is subordinate in an act of conflict to the Ambassador; we always need to keep the [military] command channel up the military side."[2] A Defense official noted that this sentiment was pervasive, recounting feedback he consistently received from mid-grade military officers who claimed that uniformed personnel "could never report to civilians—it was not allowed, and had never been done."[3]

Accompanying conceptual conservatism are concerns about the desirability of a more tightly integrated chain of command. One DoD official noted the safeguards that redundant chains provided. She maintained that unified authority was "theoretically" desirable, "but what if we chose the wrong chain of command? Now we hedge our bets with multiple lines of funding."[4] A senior FSO thought that unified authority would work if all team members had common training and cultural awareness, but noted that absent this integration of training and culture, it was best to employ split chains, leaving that "theoretical ambiguity to make sure that the military commander, with all his advantages, must at least take account of the theoretical independence of the State partner, and take into account the consequences if he runs the State official over."[5] Independent counsel was cited by a State PRT member as an unambiguous advantage of segregated command—"On a personal level, the State and USAID persons are the only ones not in the direct chain [of the PRT commander], so the commander can bounce ideas off them. While the commander can't come across as uncertain [to military personnel], he can consult these two."[6]

THE AMELIORATIVE EFFECTS OF UNIFIED AUTHORITY

Eliminating Remote Management.

The concerns in the previous section are legitimate, but do not outweigh the clear benefits of integrated command. Unified authority addresses two problems associated with remote management. First, a consolidated chain would solve the problem of underperformance isolated from supervision—now, distant oversight spawns three loosely accountable fiefdoms. Not all senior PRT officials will always perform well, and those who do not are unlikely to report their own failures objectively. The current recourse for the remainder of the team is always unpleasant, trying, inefficient, and often ineffective. Initially, one of the tri-leaders must become sufficiently frustrated to pass the complaint up his own chain of command. After arriving in Kabul, it must go back down the chain of the charged official; back up comes a response; and ideally a resolution travels down the chains of both the aggrieved and the accused. At each stage, the grievance is likely to lose urgency, and only one person need neglect it for no resolution. Additionally, as described in the "Difficulty of Distant Command" section of Section 3, senior officials in Kabul often would rather preserve their own daily relationship with their counterparts than advocate for a remote subordinate who is unable to resolve an issue locally. Because of the daunting recourse mechanism, problems often fester, or are addressed by an ad hoc solution that isolates the impact of the difficult player. This suboptimal solution is the only available choice to a team suffering poor performance by one member but without a single leader vested with the authority to correct it. Unified authority would eliminate this convoluted system; moreover, disputants, aware of a clear system for on-site resolution, would likely resolve an argument before it rose to the team leader.

Second, unified authority would improve the response time of the team. One PRT veteran noted that a team commander with real authority could, upon "realizing that day that something bad was happening, immediately swing the team members to that problem without waiting for the individual players to get permission" from distant authorities.[7] This inability to act decisively spreads beyond the team; for example, a Regional Commander might like to accommodate the PRT military leader's request to steer national agricultural programs (managed by the mission in Kabul) away from a warlord who siphons off funds and refuses to cooperate with the Afghan National Army. This request, even if accepted by all departmental leaders at the team and RC level, must first travel to USAID in Kabul, following the same complex path that characterized intrateam dispute resolution. This case also illustrates the impediments caused by the segregation of national and local USAID programs, analogous to the separate chains for the maneuver battalions and PRTs.

Diminishing the Role of Personality—Lifting the Floor, Not Raising the Roof.

General Barno cites his close relationship with Ambassador Khalilzad as crucial, and he is right that "personalities matter regardless of the structure."[8] But personalities become especially important when there is no clear responsibility and corresponding authority. Because there is no system for resolving disagreement other than parallel recourse to Kabul, PRT members waste time and energy framing the problem, estimating each other's commitment to a stated position, predicting the importance of the desired result, and comparing the stakes of the current dispute with those of possible future arguments. In anticipation of these struggles and in the absence of an integrated management system, departmental leaders must continually defend their prerogatives.

The assessment of status may begin immediately when a person arrives at the outpost, as the dynamic established in the first couple weeks has inordinate influence on turf issues. One FSO who worked at two PRTs noted that when he reported at each, "I had to make it clear to each PRT commander that I did not work for them—'My boss is the Ambassador.' . . . I got in arguments with the XO and the staff, who wondered where I fit in the military hierarchy; the XO thought I was part of the staff, and therefore worked for him. I told him that 'I don't work for the PRT commander and I sure don't work for you.'"[9] But as the management structure is ill-defined, and civilians and military may be on different rotations, the skirmishes for status continue, with petty issues evoking disproportionate passion. One USAID representative recounted returning after leave to find a recently arrived FSO in her office, claiming precedence as USAID was part of the State Department. Determined not to lose face, the USAID representative insisted to the interloper and the military commander that she keep the room, and they assented.[10] A meeker USAID officer might have capitulated, signaling weakness and yielding future advantage to State and military counterparts.

To some extent, this jousting for influence happens at any PRT, no matter the character of its members. When personalities mesh, tension recedes and departmental leaders are able to coordinate activities constructively. At the best PRTs, the three officials will create an ad hoc management structure, such as an executive committee which meets nightly to coordinate activities and hosts weekly meetings to reach out to other organizations in the province. In these cases when the colonel, the FSO, and the USAID representative work well together, unified authority might add clarity but would probably not significantly improve an already smoothly functioning team.

But when one member does not have the same vision as the others and personalities clash, the outlier, subject only to ineffective remote supervision, may simply opt out of coordination, and work the programs in his sphere with little regard for the concerns of his colleagues. Unified authority addresses these situations, creating an arbiter on location.

Perhaps more importantly, unified authority continually facilitates decision even when there is no open dispute. With no final arbiter, departmental leaders must calculate whether or not to raise an issue—is it worth the discussion, do I want to deal with the potential disagreement, will it waste my capital? Thus, many issues more minor but with significant impact on the PRT mission may not even rise to the negotiation phase, as a departmental leader accepts an undesirable status quo in order to avoid a taxing conflict.

COUPLING AUTHORITY WITH RESPONSIBILITY

The most important benefit of unified authority is the wedding of the power to act with responsibility for the actions' effects. Currently, diffuse team leadership creates the expectation that departmental representatives should stay in their own lanes,—the State FSO focuses on local diplomatic initiatives, the USAID representative on development, and the military leader on security. This bureaucratic tendency is enforced by the guidance from the NSC's Deputies Committee in 2003. This model of stovepipes may accord with American departmental divisions, but it does not accommodate the complex nature of COIN, where an action in one sector may have an immediate impact on all three.[11] One exasperated Defense official contended that "when you say 'de-conflicting,' it means that you want to stay in your lane. But guess what—your lane is changing."[12]

Collocation improves the efforts of diverse departments, as each learns the others' culture and gains empathy for the others' challenges. However, segregated chains of command dampen the possible gains. On a team, when a disagreement arises between the three leaders over the effects of an action, the two other leaders defer to the representative of the relevant department—each is responsible for his own lane. In general this is appropriate—with a career in the affected field, the expert probably has the best idea of the correct course. Moreover, most decisions in these fields are uncontroversial—USAID rep, FSO, and military officer will usually agree that a particular plan is best for a new school project or new power-sharing agreement in the provincial government.

However, when decisions are controversial, it is usually due to conflicting priorities—the military officer believes that the immediate deployment of Afghan National Police from another region will increase village security, but the FSO worries that it will undermine recent political progress made with a warlord who hoped that members of his tribe would provide manpower for a local security force. Of course, both are probably right, and each might admit the conflicting effects—the action will probably simultaneously promote immediate security and undermine negotiations. So the disagreement is not over the effects of the action, but rather the prioritization. What is the surest route to victory: an immediate drop in violence or the prospect of co-opting the warlord?

Under stovepiped chains of command, as policy-makers have not integrated authority between the departments, the decision whether to act usually rests with the team member who owns the relevant assets. As a distant supervisor, rather than an empowered PRT leader, evaluates a departmental representative, he or she will follow the priorities of his department measured by its particular metrics. This dynamic will tend to favor action over restraint—self-evaluative reports are more impressive if they demonstrate tangible results (number of security forces trained, money spent, meetings held), no matter their effect (which may be immensurable), than if they surmise misfortunes averted because of wise deferral. Thus in the debate between one departmental head who favors a particular action and a counterpart who prefers restraint, the owner of the assets will nearly always prevail, and the decision will often be to act, even in cases where inaction might produce greater net gain across all three sectors.

The original PRT mission envisioned three spheres—reconstruction (and economic development) falling to USAID, security to the military, and the extension of the authority and presence of the central government (and the accompanying facilitation of local politics) to the State Department. However, like a rock thrown in a pond, every action in COIN affects not only its own immediate sphere, but the remaining spheres as well. Segregated authorities promote a parochial perception of the war—I'll do my part without your interference, you do yours and I won't meddle, and everything will be peachy. Each department has responsibility for its programs, but no one has responsibility for their effects. This disjunct contradicts what is necessary, and what the government literature recognizes is necessary—a truly integrated program which facilitates action when appropriate and restraint when not.

A unified authority would allow true integration. Rather than three separate leaders, each biased in behalf of the positive effect of his own action in his own sphere and less concerned about its effect on other areas, a single PRT leader would have the authority to lead all action, no matter the sector, and be accountable for all effects in every sphere. Such a dynamic would be a significant improvement over the current management model in which effects, but not accountability, extend beyond the originating area of action. Decisionmaking might still suffer from the American bias toward mensurable action. But the separation of authority to act from responsibility for effects would no longer exacerbate this tendency.

ENDNOTES - SECTION VII

1. For more, see John Nagl, *Learning to Eat Soup with a Knife: Counterinsurgency Lessons from Malaya and Vietnam* , Chicago, IL: University of Chicago Press, 2005; and Andrew Krepinevich, *The Army and Vietnam*, Baltimore, MD: The Johns Hopkins University Press, 1986.

2. David Barno in discussion with author, July 18, 2007.

3. DoD official in discussion with author, July 6, 2007.

4. DoD official in discussion with author, July 9, 2007.

5. Senior FSO discussion with author, July 26, 2007.

6. PRT FSO in discussion with author, August 1, 2007.

7. State Department official in discussion with author, July 13, 2007.

8. David Barno in discussion with author, July 18, 2007.

9. FSO in discussion with author, August 1, 2007.

10. AID Representative in discussion with author, July 17, 2007.

11. Nor does the tripartite model fully capture the complexity of insurgency which includes more than simply three spheres; however, if interpreted broadly, these three areas are sufficient for a basic system.

12. DoD official in discussion with author, July 8, 2007.

SECTION VIII. A NEW STRUCTURE

As in the first half of the Vietnam War, a multitrack chain of command has plagued efforts in Afghanistan. The administration should impose unified authority over the American efforts, designing an organization to support a strategy, not vice versa. As in Vietnam, the government should create a new hybrid organization, comprised of employees of deployed departments, dedicated solely to guiding the American effort in Afghanistan, and enduring only as long as the war. This new entity would focus participants on the fight, rather than on the interests of their parent departments. The chain of command should reflect the political strategy, operations, and tactics of a COIN campaign. Following the CORDS model at the tactical level, Regional Commands and PRTs should be directed by either a civilian (from USAID or State) with a uniformed deputy, or a military officer with a civilian deputy, as determined by the level of violence in the area of responsibility; the commander would be responsible for the performance within the entire assigned geographic area, and would have requisite authority over all team members, no matter their home agency. Furthermore, PRTs should subsume maneuver units, bringing them under the civil-military team commander; a military officer should invariably lead provinces where strike groups still operate. At the theater level, all American efforts, civilian and military, should be unified under a civilian Chief of Mission with real authority, aided by a military deputy. Considering the growing role of ISAF, all operations in Afghanistan, no matter the responsible country, should fall under the authority of a senior NATO Ambassador, with a senior NATO general as deputy.

MODEL

The Country Team.

COIN demands meticulous civilian guidance as forces negotiate the political intricacies of the host country, at a local as well as national level—a civilian is best equipped to provide this direction, and furthermore shows the population a model of military subordination to political control. Additionally, civilian leadership would facilitate the access of the various departments to military resources. Finally, recognizing that the qualities of the individual leader have a tremendous impact on the effectiveness of the campaign, the use of a presidential representative of ambassadorial rank allows the President to choose from all possible candidates, civilian, and recently retired military.

In Afghanistan, a civilian proconsul should guide all American efforts. War is always an extension of politics in its macro effect, and insurgency is political at the micro level as well. No matter the military might, the effectiveness of COIN is still constrained by political goals. These goals are linked to the means that the Afghan government is willing to expend and the compromises opponents and partners are willing to make. In general, the State Department is better suited than DoD to handle the vagaries of negotiations that guide political efforts in Kabul and the provinces. Additionally, ambassadorial leadership of all American efforts will give a civilian face to the host nation, signaling to the young Afghan government the importance of civilian control of the military.

In addition to aligning practice with theory, civilian leadership of a unified civil-military country team would have two practical benefits. First, the military will continue to play a very large role in Afghanistan due to its advantages of resources over the civilian agencies. Until Congress closes the gap in resources, which is unlikely in the near future, placing a civilian at the head of the integrated operational chain of command will have the same effect that Komer achieved in Vietnam (or perhaps greater, as Komer was still subordinate to the Commander of MACV) —a dramatically increased availability of military resources to the civilian departments.

Moreover, this system will give a President more flexibility in choosing his proconsul. The path to flag rank is fairly rigid in peacetime, and recently the executive has been reluctant to interfere in this military process. However, the President always appoints ambassadors. Furthermore, the nominee is often found outside the State Department. By placing the ambassador at the head of the Country Team, the President may pick whatever American is best-suited for the particular challenges—a person who has proven not only to have substantive expertise, but more importantly is an outstanding leader who understands the exigencies of war as well as Afghan culture. Whatever the professional origins, the President will grant the ambassador the authority to control all American assets in Afghanistan, and hold him accountable for progress.

If an ambassador is to assume this role, the country team staff must expand dramatically to accommodate the increased responsibility. General Barno recognized this when he seconded some of his military staff to Ambassador Khalilzad. One FSO who served at a PRT noted that currently in Afghanistan, the departmental representatives on the Country Team served not as a staff but rather as a looser "council of advisers . . . unable to get anything done. Even if they all agree on something, there is no implementation process."[1] An invigorated Chief of Mission (CoM) requires a robust staff so that he can plan, implement, direct, and monitor all American programs, civilian and military, in Afghanistan.

The Tactical Level.

From the country team down to the tactical level, a single chain should guide all action. A mix of military commanders, with a civilian deputy, or senior diplomats, with an O-6 executive officer, should lead the Regional Commands. Reporting to the Regional Commands, PRT commanders should be military with a civilian deputy and vice versa. In stable provinces with no strike units present, the civilian should lead; as is the case with the CoM, this will present an example of civilian supremacy over the military to the provincial government.

No maneuver units should escape tactical integration. General Barno extended the principle of geographic responsibility to the Regional Commands; his logic should apply to the provincial level. A uniformed PRT leader would supervise maneuver units to ensure that their operations did not undermine the PRT's broader mission. This shift to the PRT would accelerate a cultural shift within the military, elevating command of a team to the most coveted position available to a colonel, thereby encouraging the conception of command as a holistic, not simply kinetic, duty. This subordination of the maneuver unit to the PRT is particularly suited for the intensity of the war. It is rare that insurgents coalesce in company-sized bands. Army planners might split the current battalions into maneuver companies. When accompanied by American airpower, a company will usually be sufficiently powerful to defeat any large insurgent units. Should circumstances demand a larger group, the PRT commander or Regional Commander could organize a contingency joint task force, comprised of maneuver companies contributed by one or, if insufficient, several PRTs.

This proposal raises concerns that reconstruction might be conflated with the strike mission. Many claim that locals are more comfortable with PRT personnel than tactical units because the teams are able to disassociate themselves from the tough tactics of the maneuver battalions. These concerns are legitimate. However, under this new structure, maneuver elements would work for a PRT leader, responsible for the entire American portion of the COIN campaign in the province. Afforded real authority, this leader could modulate the tactics employed by a strike unit, reducing the gap between its approach to the population and that of the team.

ISAF.

This paper has focused on American efforts in Afghanistan. Increasingly, other ISAF nations bear the burden of the campaign; today, nearly half the troops hail from coalition partners. Most of the American military effort now, appropriately, falls under NATO.

Because of cultural differences between nations and the difficulty each country has in coordinating its own varied efforts, American forces should integrate with ISAF at the level of high command, but not the tactical level. The United States currently commands the East Region, which demands the most resources and personnel; the remaining four fall under Italian, British, German, and Turkish commands.[2] In this new model, RC responsibility would be proportional to personnel contributions, giving the United States a second RC to accompany the eastern region. The proconsular Chief of Mission would direct all American forces, civilian and military, within the U.S. RCs, and all American resources would be focused in these regions. Ideally, Congress would fund the American effort through a war budget, rather than within individual departmental budgets, giving this ambassador broad financial control. Other ISAF contributors would provide military as well as civilian reconstruction and diplomatic personnel to their regions of responsibility.

Should American personnel offer a unique capability needed outside of the American regions, these units, whether civilian or military, should be put under the operational control of the relevant PRT,[3] whatever its nationality. However, since the cultural gulf is truly wide between different national governments, this should be done only when absolutely necessary. A senior NATO Ambassador should preside over ISAF efforts; reporting to him should be a Senior NATO general. As long as the United States contributes half of the forces, an American should hold one of these two positions. If other nations adopted a model similar to the proposal for the United States, the RCs could report directly to ISAF command which would direct all civilian and military programs, eliminating the need for an American proconsul. As this is unlikely, a powerful ambassador will be needed to direct the two American RCs.

This coalition management arrangement, focused on the regions, would eliminate many pan-Afghanistan programs (with a few exceptions, such as the ring road, that must be managed by a single party no matter what regions are involved), accepting that cost for the benefit of integrated spheres at the local and regional levels. At the coalition level, this system is less than optimal,[4] creating potential for a NATO member to finance local projects in its own geographic area that are difficult for the ISAF commander to coordinate with projects run by other nations in the neighboring regions (the ISAF commander has only tenuous influence over each country's reconstruction programs). But these inefficiencies are inevitable in coalition warfare with different sovereigns. At the local and regional levels, there is no need to incur such a cost for American multiagency operations.

Achieving the New Model.

Practically, how might this integrated model operate? It will not give the leaders of the new organization the same power as military command—for example, the civilian leader could not mete out nonjudicial punishment. But it will certainly give each, from ambassador to team leader, the same level of authority over the other departmental representatives as exercised within a civilian organization. This includes the power to direct activities, the responsibility for managing subordinates (including hiring, firing, and evaluation), and control of the integrated funding that had previously been divided along departmental lines. The new organization should also have substantial influence over deployment schedules.

An effective leader will rely on subordinates for the execution of tasking within their spheres. Furthermore, the leader will be responsible for all the operations of the lieutenants—in this case all American security, political, and development activities at the provincial, regional, or mission-wide level. One arbiter will be able to balance the competing priorities of each sector's advocate with the broader effects of single actions across the spectrum.

AID and State representatives do not have as many leadership opportunities as their military counterparts. But this does not mean that neither agency has any good leaders. The number of such authority positions open to USAID and State would be low, perhaps 6-8 PRTs and one RC; each organization would need to find fewer than five good leaders. The mission will attract enterprising candidates, many with military experience who crave a return to leadership. To complement this self-selection, the departments should create career incentives, such as accelerated promotion, to reward those who serve PRT tours. Additionally, they could advertise PRT duty as the most challenging, and noble service. With this combination, State and USAID would recruit superior volunteers to fill allotted positions.

The most difficult dynamic will be civilian leadership of uniformed members on a PRT—without military training, experience in giving orders, and familiarity with the mortal consequences of combat, how might USAID and State officials direct soldiers? Selection processes and a formalized command relationship would help overcome these obstacles. A board for the new organization, comprising members of every participating department, should select the team leaders—criteria should always include leadership and usually military experience. Additionally, explicit command guidelines might mandate that while a civilian team leader can order or cancel operations, he should leave the tactical guidance of the operations to his military XO. Finally, civilians will lead teams only in stable provinces, with no maneuver units.

CODA

The lack of unified authority is not the only hindrance to our COIN effort in Afghanistan. The chief challenges are the inability of the Afghan government to provide for and control its people; the tenacity, diversity, and adaptability of the enemy; and the Pakistan haven for insurgents. Other obstacles include the distance between coalition personnel and Afghan officials, low coalition manning, organizations ill-suited for expeditionary work, and a poor understanding of Afghan history, culture, and languages. Coalition warfare brings more problems—restrictive national rules of employment limit field work, fickle domestic politics induces operational conservatism if not paralysis, and cultural and doctrinal differences between allies inhibit cooperation. Nor is absence of a clear command system the only impediment to improved American country team or PRT performance. Integration per se will not overcome the shortage of civilian staff and insufficient number of teams (25 provincial teams as compared with 44 provincial and 250 district teams in Vietnam). I have focused on the chain of command because, unlike many of the other issues, its reform would demand almost no additional funding, the change would require little extra (and only for the CoM's staff) in manning, and its effects on our COIN efforts would be immediate.

PRTs are the centerpiece of our COIN efforts in Afghanistan. Yet no one in the field knows what single leader or organization is responsible for the program's performance. The same lack of unitary accountability mars our nationwide programs. This problem is especially galling since the U.S. government has not even begun to replicate the half-decade reform effort that led to CORDS and has ignored its own warfighting principles so clearly expounded in Goldwater-Nichols and the Joint publications. What is surprising is not that the bureaucracies, eager to protect their prerogatives, should resist integration, but that the government should hope to produce the effect of unified authority (i.e. unity of effort) without employing the proven mechanism.

PRTs are not without success. They have contributed significantly to the extension of Kabul's power into the provinces. Nor is the current structure without merit—collocation has brought departments together at the tactical level, which is bound to increase the level of coordination between members. But even in the best functioning team with conscientious personnel, the maintenance of departmental stovepipes separates authority for action from responsibility for effects and may undermine ad hoc attempts at integration. When personalities do not mesh, the result is much worse, as one team member, safely distant from oversight, may opt out of operations. These lessons in some ways apply even more strongly to our regional and national efforts, which are not collocated like PRTs and lose out on the corresponding benefits.

Some propose a consensus model for multiagency command. Consensus is a leadership style, and can be very effective. Involved in the creation of a program, subordinates feel invested in its execution and remain motivated through its duration. However, consensus is not a system of management. An organization dependent on agreement from every member for decision will either drift towards paralysis or issue vapid decrees. Inevitably, each participant has a different conception of effectiveness, and the only way to achieve complete agreement is to create a product that is worse than the sum of the individual parts—one that, in the case of the U.S. Government, preserves the sacred elements of each individual bureaucracy in order to buy assent. While none dispute the need for integrated response, each institution's most important goal is usually to preserve an area of expertise or other turf for itself—this exacerbates the natural tendency of fragmented responses to a challenge.

I have proposed a CORDS-style command structure, with some important enhancements—an organization that has the authority to direct all American activities in Afghanistan, no matter the parent department, and has responsibility for all effects. As declared in JP 1, unified authority will create true unity of effort horizontally (across the departments) and vertically (from municipal to national programs). The proposed structure puts the ambassador in charge of all efforts, with authority delegated to leaders of the three primary expeditionary departments who will direct the RCs and PRTs. Authority and accountability will rest in the same chain. This structure gives flexibility to the hybrid organization, allowing the appointment of the most appropriate professional for the demands of the particular province, region, and country.

This paper offers one proposal, not the only possibility. There are practical reasons for skewing authority towards civilians. But any arrangement of concentrated authority in an able and eager manager with real control over all aspects of the effort is better than the current ambiguity. Put someone in charge and demand leadership.

ENDNOTES - SECTION VIII

1. FSO discussion with author, July 24, 2007.

2. See *www.nato.int/isaf/structure/regional_command/index.html*.

3. This assumes that the sponsor nation reciprocates.

4. Assuming that coalition members do not cede all authority to ISAF, they will retain national restrictions and control over their assistance programs.

APPENDIX I

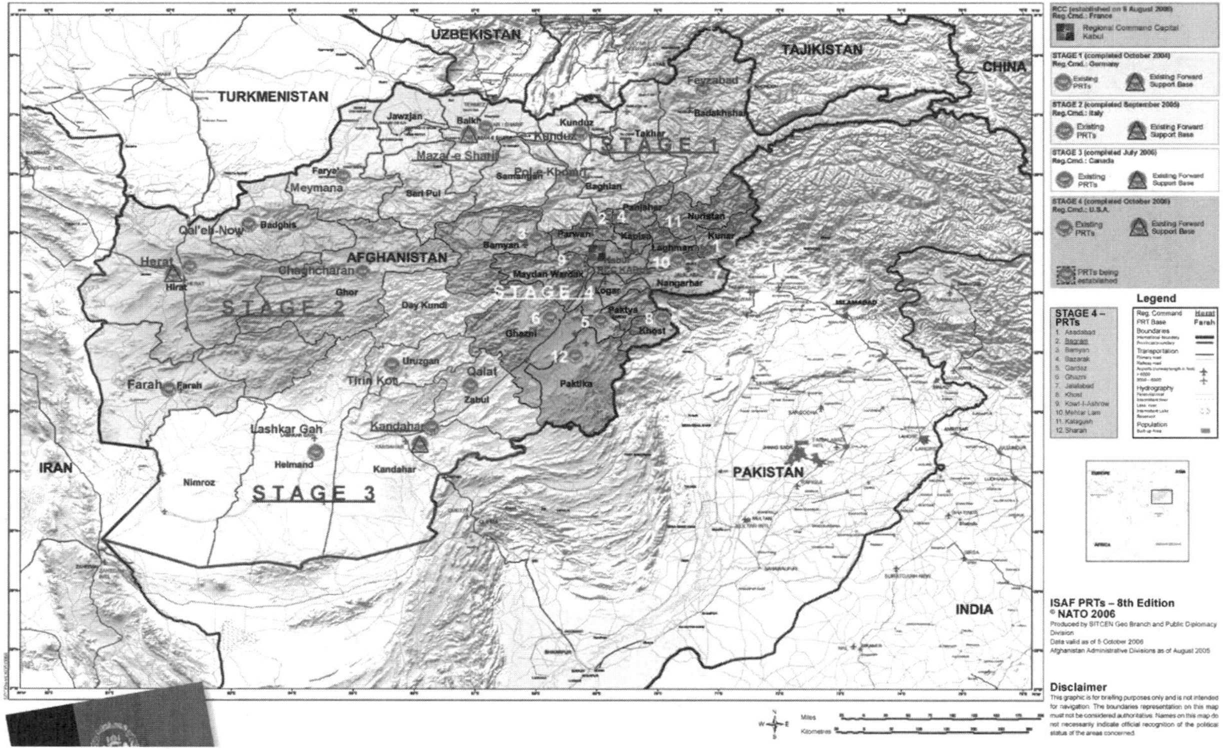

From NATO/OTAN Afghanistan Briefing, 2006, available at *www.nato.int/docu/briefing/afghanistan-2006/afghanistan_e.pdf*.

APPENDIX II

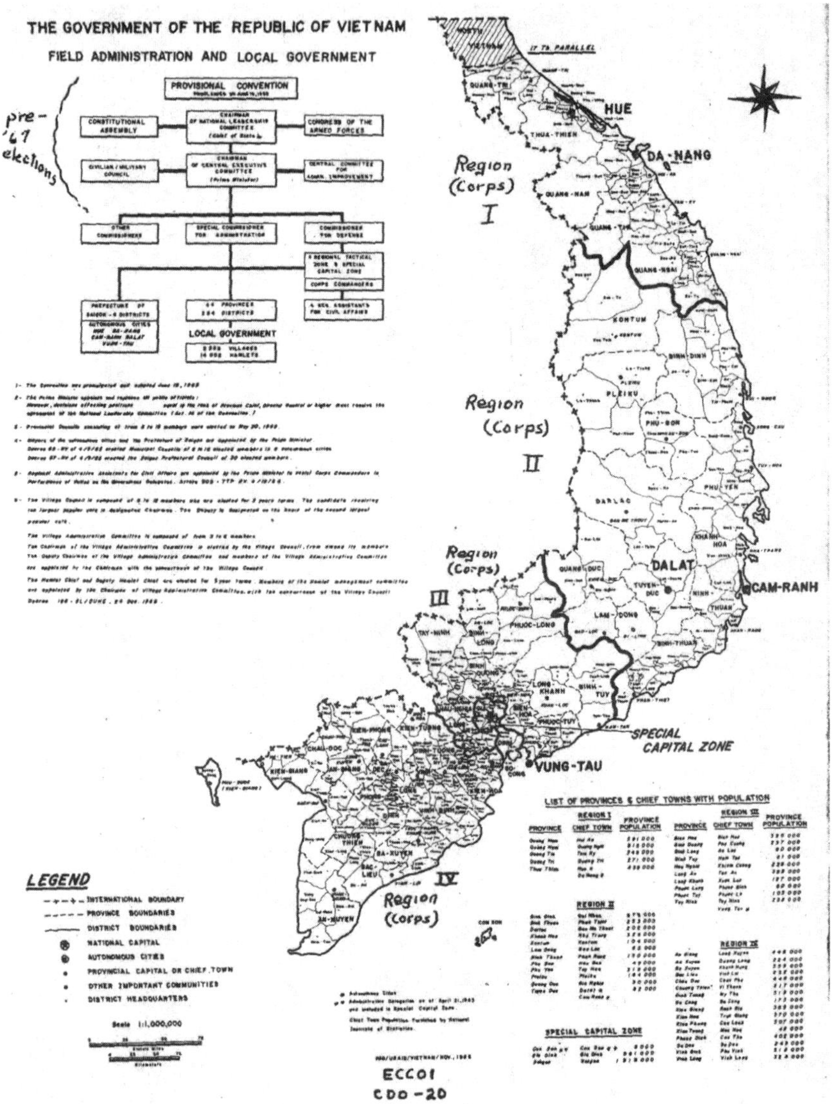

Image from T. Malupit, ed., *Vietnam and Counter-Insurgency,* Quezon City, Philippines: Eastern Construction Company, Inc., 1967, p. 153.

APPENDIX III

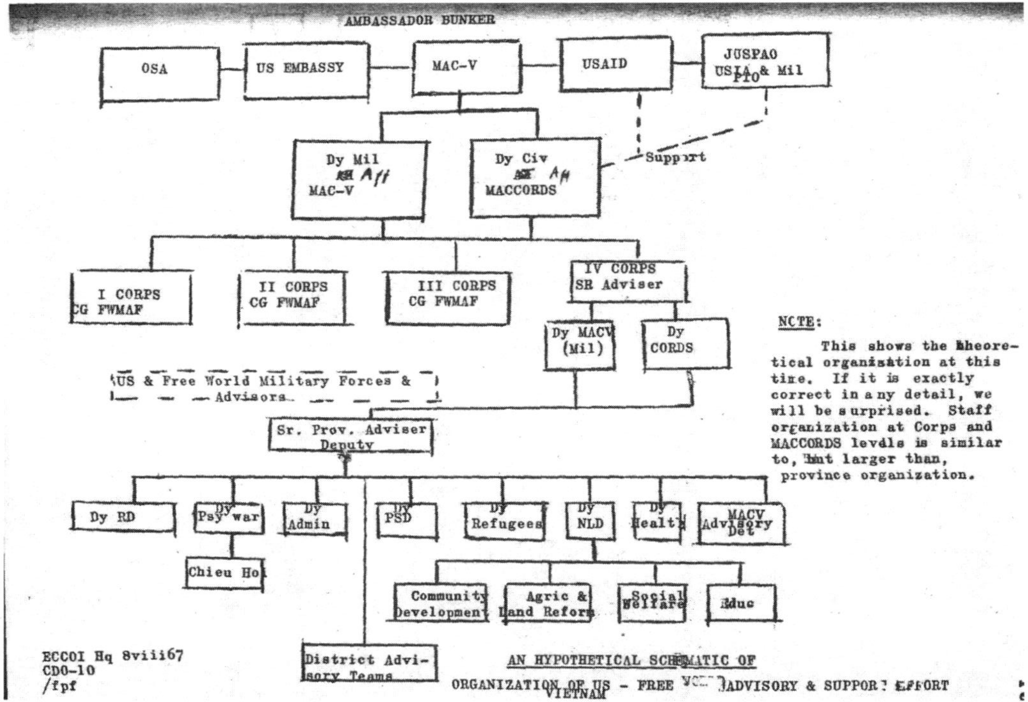

Image from T. Malupit, ed., *Vietnam and Counter-Insurgency,* Quezon City, Philippines: Eastern Construction Company, Inc., 1967, pp. IX-67.

APPENDIX IV

Acronyms

AID – see "USAID"
AO – Area of operations
AOR – Area of responsibility
ARVN – Army of South Vietnam
CA – Civil Affairs
CCDR – Combatant Commander
CENTCOM – U.S. Central Command
CERP – Commander's Emergency Response Program
CFC-A – Combined Forces Command-Afghanistan
CIA – Central Intelligence Agency
CJCMOTF – Coalition Joint Civil-Military Operations Task Force
CJTF-180 – Combined Joint Task Force 180
COIN – Counterinsurgency
COM – Chief of Mission
COMUSMACV – Commander, U.S. Military Assistance Command-Vietnam
CORDS – Civil Operations and Revolutionary Development Support
CSTC-A – Commander, Security Training Command-Afghanistan
DEPCORDS – Deputy for CORDS
DOD – U.S. Department of Defense
DOS – U.S. Department of State
DPSA – Deputy Provincial Senior Adivisor
FSO – Foreign Service Officer
GS – General schedule (U.S. government seniority scale)
GVN – Government of South Vietnam

GWOT – Global War on Terror
HQ – Headquarters
IED – Improvised Explosive Device
IGO – Intergovernmental organization
ISAF – International Security Assistance Force
JFC – Joint Force Commander
JP – Joint Publication
JSOTF – Joint Special Operations Task Force
MACV – Military Assistance Command-Vietnam
MOI – Ministry of Interior
MP – Military Police
NATO – North Atlantic Treaty Organization
NCO – Non-commissioned officer
NGO – Non-governmental organization
NLD – New Life Development
NSPD – National Security Presidential Directive
NVA – North Vietnamese Army
O-5 – Lieutenant Colonel rank in U.S. military
O-6 – Colonel rank in U.S. military
OCO – Office of Civil Operations
OHDACA – Overseas Humanitarian Disaster Assistance and Civil Aid
POLAD – Political Advisor
PROVN – Program for the Pacification and Long-Term Development of South Vietnam
PRT – Provincial Reconstruction Team
PSA – Provincial Senior Advisor
QIP – Quick Impact Project
RC – Regional Command
RDS – Revolutionary Development Support
RF/PF – Regular Forces and Popular Forces
RMA – Revolution in military affairs
SOCOM – Special Operations Command
UCMJ – Uniform Code of Military Justice

UN – United Nations
UNAMA – U.N. Assistance Mission to Afghanistan
USAID – U.S. Agency for International Development
USDA – U.S. Department of Agriculture
USIA – U.S. Information Agency
VC – Viet Cong
XO – Executive Officer